Copyright © 2022 Prof

All rights reserved.

ISBN: 9798425375308

PREFACE

This book is a collection of tweets from Team Reality…

I came up with the idea of this book first as a printed keepsake that would showcase and honor the members of Team Reality, who have spent day and night for two years fighting the pandemic of the uninformed. Something that would stay around for much longer than our Twitter profiles. My initial intention was organizing the book by people, with each person getting their own pages in which I would post a few tweets from them showing their contribution to the fight.

However, as I began curating the tweets, I realized how good of a story they were telling. They were telling the story of how massively the "experts" had failed, over and over and over, but were never held to account. They were telling the story of how the society lost all perception of risk and fell victims to hysteria constantly pumped by said "experts" and the media. They were telling the story of how lockdowns and school closures were policies with catastrophic consequences yet were constantly pushed by said group of "experts". They were telling the story of how COVID posed no risk to most people, but this fact was never acknowledged by the "experts". They were telling these stories convincingly with data and charts and resources, instead of with emotional pleas and scare tactics. I thought this was too good of a story to not tell.

As a result, I digressed from my original plan and decided to organize the book by month, and put prominent and most relevant tweets from Team Reality members under each month. This would serve two purposes. First, it would serve as a historical account of

Team Reality

Fighting the Pandemic of the Uninformed

By Prof (@covidtweets)

R(g) RATIONAL GROUND

the fight and honor those who participated in it. Second, it would show how the pandemic unfolded, but not from the perspective of the mainstream media, which was what most people were exposed to, but from the perspective of Team Reality. I tried to make sure there is at least one tweet in this book (often more) to serve as a reminder of every single blunder that the "experts" committed, and Team Reality was always there to call them out. I believe it is crucial that these tweets are documented and immortalized, as they show how even in the darkest moment, there were those who saw through the hysteria and never lost sight of the reality. There were people who shone their brightest in that darkest moment to enlighten those who were open-minded enough to look. Those who never forgot that there were costs to "15 days to slow the spread" and school closures and most other policies which were presented to us.

After I changed my plans regarding the organization of the book, I considered to have a section in which I would add bios of those whose tweets were included. Then I decided not to do that either. Even though I tried to include as many people as possible from Team Reality in this book, there were those who had to pass for personal or professional reasons. There were also those who never check their DMs (apparently ☺). I didn't want to exclude them by emphasizing the people too much. In this book or not, every person who fought the insanity that unfolded in March 2020 and beyond is a member of Team Reality. Some just happened to contribute more, were checking their DMs, and were able to say yes when I asked them if I could use their tweets for this book.

As you will see, I also tried to make sure each view on important subjects is reflected in this book. While mostly in agreement on the major topics, in fact Team Reality is a fairly heterogenous bunch. Some wouldn't even identify as Team Reality members (although we think they are). Some are more centrist while others are more on the extremes of certain issues. They don't all agree with each other on everything. However, they all share the very important ability of critical thinking, which allowed them to see that something was very wrong. Most came to the same conclusions with regards to what was wrong, but there is occasionally some disagreement on how to make it right. Which is fine. They always engaged, with both each other and with those they were trying to "wake up", with civility and respect. Unlike a certain group of "experts" and their followers who almost always insulted and then blocked us…

One final point: The tweets in this book represent the current (at the time the tweets were posted – Team Reality members have this rare trait that they change their positions if data begin suggesting otherwise) understanding of how COVID and policies work, based on available studies and data as interpreted by those who contributed their tweets to this book. However, most team reality members are not licensed medical professionals. Even those who are would likely not want their tweets to be taken as medical advice. Neither now, nor (more importantly) in the future. Therefore, if you are reading this book, please know that the tweets in this book are not intended to be used as medical advice in any way. For those matters, please see a licensed medical professional.

This book is dedicated to the people who were harmed not by COVID, but by the humanity's response to it...

People who lost their livelihoods, their life savings, their businesses. People who lost their mental health because they were told that other humans were dangerous. People who lost their physical health because they missed their routine care. People who lost their physical health because they were made to believe it was too dangerous to leave their houses...

Our young ones, who lost precious years from their lives, who were starved of a normal childhood. Who were not allowed in their schools and were instead told that looking at a screen for hours was school. Who longed to see a smiling face from their teachers and friends, but instead had to look at half-faces all day long. Teenagers whose dates and sports were taken away, who had to have Zoom graduations or proms. College students who were forced to pay outrageous amounts of money and then be forced to live in prison cells...

In the hands of incompetent "experts", humanity suffered. In the hands of clueless governors, constituents suffered. In the hands of clickbait and social media, consumers of information fell victim to hysteria...

Never again.

TEAM REALITY

ACKNOWLEDGMENTS

I would like to thank members of Team Reality who allowed me to use their tweets for this project. You should not be forgotten, and this book will immortalize your fight.

I also thank my wife and kids, who were the reason I joined Team Reality in the first place. I apologize to my wife for spending so much time on Twitter over the past two years, but she knows it was important…

TEAM REALITY

1. INTRODUCTION

In March 2020, the world came to a standstill…

What began as scattered social media accounts of a new virus in China became the dominating story in the whole world for the next two years. Like most people in Team Reality, I was one of the early followers of what was happening in that corner of the world. In February 2020, I followed the case counts from China daily. I saw videos of men dropping dead at random places. I watched as they built massive field hospitals in a matter of days. I watched how they were "locking down" an entire city of 15 million people. I watched how they were apparently welding people into their apartments, or "disinfecting" massive cities with trucks spraying what appeared to be a white aerosolized gas into the open air.

Then, supposedly because there is a large population of Chinese workers who work in clothing factories in Northern Italy[1], who had just traveled back to Italy after Chinese New Year celebrations, Italy became the second country where the virus began spreading. Reports of overflowing hospitals in Bergamo began appearing on social media. Stories of how ventilators were not sufficient, so they were splitting them to be used on two patients at the same time[2], were going viral. Then something shocking happened: Italy, too, was following the Chinese playbook and introducing a lockdown in Bergamo…

While these were happening, the administration in the US was staying relatively calm. At first, that is. New Yorkers were being urged to visit Chinatown to shop and dine in the businesses there by their mayor. In fact, on February 13th, Speaker Corey Johnson would say, "*Unfortunately many businesses and restaurants in Chinatown, Flushing and Sunset Park are suffering because some customers are afraid of the coronavirus. But those fears are not based on facts and science. The risk of infection to New Yorkers is low. There is no need to avoid public spaces. I urge everyone to dine and shop as usual.*[4]" Pelosi was visiting San Francisco's Chinatown to show that "everything was fine"[5]. When President Trump banned travel from China at the end of January, he was being accused of being xenophobic[6], ironically by the same people who would later demand the most draconian measures and bans...

The reason why these politicians were trying to calm people's fears was that despite the governments staying sane, the people had already started to panic, and notably were avoiding areas with a high likelihood of seeing Asian people. While COVID tests were still at short supply, the numbers of cases were steadily increasing since the first case was recorded back on January 21st in Washington state. By the end of February, the total number of recorded cases would reach 70 in the USA. Still low, but people were starting to feel uneasy. Then, in early March, things began to escalate...

On March 11th, the World Health Organization declared COVID-19 a pandemic, after over 100,000 cases being recorded across the globe in over 100 countries. The next day in a press conference, NYC Mayor Bill de Blasio would say "Yesterday morning seems like a long time ago. We got a lot of information in the course of a day yesterday and a lot changed then, then last night it just seemed the world turned upside down in the course of just a few hours." The next day, President Trump declared national emergency to combat coronavirus. If I was asked to pick a time when the relatively measured policy of the early days was replaced with unhinged government over-reach, I would likely say it happened between March 11-13, 2020.

Governments panicked... Within a matter of days, every state announced some version of a "lockdown", a method that would later become the go to measure of governments around the world whenever cases of the "novel coronavirus" surged. Students were told that their schools were no longer safe, and that they would now be learning remotely from their homes. Teachers in K-12 schools and professors in colleges

scrambled to haphazardly put some online material together so they would be able to continue doing something that resembled teaching. Like dominoes, everything that made life worth living got cancelled or were shut down, one by one. Sporting events, weddings, shows... Even funerals were banned...

On March 16th, a team led by Neil Ferguson of The Imperial College came up with a model that showed that if the kept going to Chinatown, or any other business for that matter, around over two million Americans would lose their lives[7]. This did not include those who would die because the demand for ventilators would far exceed the available supply and therefore every person who would need mechanical ventilation would die, increasing the death toll to over four million. The only way to avoid this, the report said, was implementing a "suppression" strategy, which would "require a combination of social distancing of the entire population".

Also on March 16, the government was announcing their brilliant strategy to control the spread: "15 days to Slow the spread". This was the intervention we were told was needed in order to "flatten the curve", a term which would be highly popular in the early months of the pandemic. The idea was that by staying home except for essential activities, we would avoid getting infected all at the same time, which would ensure that there would be enough hospital capacity for those who need it. Even though this implied that the number of people getting infected would stay the same, only with a "flatter curve" and over a longer time period, most people were not aware of this aspect of the math at the time.

The first place to get hit hard in the US would be New York City. From the second half of March to late April, I, along with many others on Team Reality, tuned in whenever Cuomo had a press briefing. I watched as the numbers kept climbing. In the meantime, models continued to be published. The US government was following the models by The Institute for Health Metrics and Evaluation (IHME) and announcing policy decisions based on what the model predicted. Along with Cuomo's daily briefings, another sight I caught daily had become the White House briefings by President Trump, in which he would let Dr. Fauci and Dr. Birx explain the situation and share projections. I will be honest, I initially had great respect for Fauci and Birx, both of whom appeared to be calm and measured. I did not know them before the pandemic. They were explaining the situation and were not using scare tactics to get the populace do what they wanted. Although, to be fair, most of the country was terrified and were doing exactly what they were asked to do. In March and April of 2020, there was not much resistance from the populace against the insane measures taken by the government, because most people were genuinely scared.

However, people like me who were following the data from the early days had a different reaction. Even though data from China was (and still is) highly suspect, data coming from Italy and South Korea had also confirmed that the death rate was extremely low for people younger than 70, and deaths among kids were virtually non-existent. This would be a major sigh of relief for most people and also parents like myself, but unfortunately, this was not the message that was being communicated. In fact, for a very long time, maybe for as long as until well into 2021, official messaging and policy never even acknowledged the age-dependence of the risk of infection…

On March 19, an influential article on Medium called "Coronavirus: The Hammer and the Dance" argued that we needed to temporarily implement strong mitigation measures, which would drastically reduce the transmission of the virus within 3-7 weeks, after which point we would "dance", meaning we would implement sustainable measures to keep R (reproduction rate) below 1 until vaccines were available. Many people who were interested in what was happening, including myself, found some of the ideas intriguing at the time. The article was widely shared by influential figures and was even incorporated into official policy documents in some places around the world in the months that followed. However, looking back from 2022 with the knowledge I gained over the past two years, I can easily see how the ideas touted in the article were bound to fail, not because people would not care and would not follow the guidance (they did, for much longer than 3-7 weeks), but because the article was based on a very limited understanding of how the virus was spreading. In addition, it is important to note that even "Hammer and Dance" argued for "dance" until vaccines were widely available. Virtually nobody would predict that we would still be "dancing", here in 2022, one year after the vaccines were approved…

I was also catching first signs of a peak in New York as we approached the second half of April, but said models were showing something entirely different. In fact, the models' predictions were so absurd that they were wrong as soon as they were published. For example, while the IHME model would show 6,000 hospitalizations for the current day, there would be 5,000 actual hospitalizations. I emailed the modelers about the obvious discrepancy, but I never received a response. Not that I was expecting one, but I couldn't sit on the sidelines either. I had to do something. I had to contribute to the conversation.

Then I found my people on Twitter…

There was one person who was also pointing out the discrepancy in the hospitalization numbers between what IHME was putting out and what was happening on the ground. His name was Alex Berenson. I didn't know him at the time, but according to his bio, he was a former NYT reporter, and apparently, he had done some reporting and written a book on the harms of marijuana. His COVID takes appeared to be in line with my thinking. Once I began following him, I quickly discovered that I was not alone. There were many people who were also following the data and calling out the errors in the models. Over time, these people began to call themselves Team Reality (if my memory is correct, a name which was first mentioned by Alex Berenson).

The rest of the story will be told by them. Throughout all of this, they relentlessly fought against the hysteria. They were the ones whose judgment was not clouded by fear or politics. They clearly saw that many things we were asked to do to protect ourselves and those around us were simply not working but instead nothing more than knee-jerk reactions. The problem is, a knee-jerk reaction takes a moment to happen and then goes away. In this case however, the knee kept kicking long after it was obvious that there was nothing there to be kicked and we were instead kicking thin air. After a while, the knee got tired of kicking, began straining its ligaments, but the "experts" insisted that the knee couldn't stop kicking because if it did, very bad things would happen. Those in Team Reality saw through the hysteria and called it out, but when they did, they were accused of "just wanting to get their nails done" or "not caring if grandma died". They were even called "grandma killers", ironically by the same people who never accepted that the risk of death was exponentially higher for older people than it was for kids. They are the unsung heroes of this pandemic, as they gave hundreds of thousands of people hope in the darkest days of most relentless fearmongering by the "experts". This book is an attempt to document and immortalize their fight. By telling the story of the pandemic through their tweets, this book will show that as long as there is at least one person who is not afraid of calling out the wrong, others will soon join them because even though it takes courage to go against the herd, courage is contagious. This is the single most important lesson we can leave to our kids, as there is no guarantee that something like this will not happen again in the future, and when that happens, they should know that even when it feels like they are alone, they should not hesitate to speak up.

People in Team Reality are heterogenous in their views. They don't agree on everything. Some believe lockdowns may be justified under certain circumstances whereas other vehemently oppose them. Some don't mind masks as much while others see them as the most important hill to die on. Some believe COVID vaccines saved lives while others have doubts. Some are more centrist while some have more extreme views. Despite this, there is one thing that unites them: They never fell for the groupthink. They never blindly accepted the mainstream narrative. They acknowledged trade-offs. They questioned and looked at the data for themselves. When what they saw was not pretty, they spoke up. They were labelled anti-vax, conspiracy theorists, contrarians, etc. by the mobs, but they kept fighting the good fight. There are simply too many people in Team Reality to include all of them in this book. Because of this, the next section will include tweets by a number of members who had a relatively bigger role in the fight against groupthink and hysteria. While it is not possible to include every single one, I just want to say that if you fought against the insanity, not because you don't care about saving lives but exactly because you do, if you fought against the harm that was being done to those among us with the least voice, to the most vulnerable, including our working class, our kids, our parents and grandparents in nursing homes, thank you. If you saw the irony in "we are all in this together", if you saw that the whole pandemic response was a collection of restrictions which disproportionately affected those with the least amount of resources, while allowing the "Zoom class" to live their best lives from the comfort of their homes where they watched Netflix and had Doordash deliver food to them, and especially if you had the courage to speak up against this, thank you. This book is also your story. You should be proud...

2. THE TIMELINE

MARCH/APRIL 2020

Mark Changizi
@MarkChangizi

The moral of coronavirus19 will be that social contagion via social networks is more dangerous than biological contagion.

3:21 PM · Mar 17, 2020 · Twitter Web App

TEAM REALITY

Prof François Balloux ✓
@BallouxFrancois

I should be qualified to comment on the covid-19 pandemic. I'm a computational/system biologist working on infectious diseases and have spent five years in a world class 'pandemic response modelling' unit. In this thread, I will summarise what I believe I (don't) know. (1/12)

10:39 AM · Mar 14, 2020 · Twitter Web App

After having spent considerable time thinking how to mitigate and manage this pandemic, and analysing the available data. I failed to identify the best course of action. Even worse, I'm not sure there is such a thing as an acceptable solution to the problem we are facing. ++ I believe that the covid-19 pandemic is the most serious global public health threat humanity faced since the 1918/19 influenza pandemic. There are major differences between the two events but I suspect there will also be similarities that may emerge once we look back. ++ The most plausible scenario to me is for the covid-19 pandemic to wane in the late spring (in the Northern hemisphere), and come back as a second wave in the winter, which I expect could be even worse than what we're facing now. Pic below is what happened in 1918/19. ++ Predictions from any model are only as good as the data that parametrised it. There are two major unknowns at this stage. (1) We don't know to what extent covid-19 transmission will be seasonal. (2) We don't know if covid-19 infection induces long-lasting immunity. ++ Seasonality is difficult to predict without time-series. Comparison between regions for the covid-19 pandemic suggests some seasonality, but likely less than for influenza. This would be roughly in line with other Coronaviridae (common cold and MERS). ++ How long immunity lasts for following covid-19 infection is the biggest unknown. Comparison with other Coronaviridae suggests it may be relatively short-lived (i.e. months). If this were to be confirmed, it would add to the challenge of managing the pandemic. ++ Short-lived immunisation would defeat both 'flattening the curve' and 'herd immunity' approaches. Devising an effective strategy would be even more challenging under low seasonal forcing. It would also considerably complicate effective vaccination campaigns. ++ The covid-19 pandemic is an extremely challenging problem and there are still many unknowns. There is no simple fix, and poorly thought-out interventions could make the situation even worse, massively so. ++ The covid-19 pandemic is not just an epidemiological problem. It is a 'Global Health' problem, that can only be tackled with an integrated and global approach. For example, there is no such thing as a choice between managing the pandemic vs. protecting the economy. ++ Health and the economy are closely linked. The correlation between per-capita GDP and health (life expectancy) is essentially perfect. If the covid-19 pandemic leads to a global economy collapse, many more lives will be lost than covid-19 would ever be able to claim.

Bethany S. Mandel ✓
@bethanyshondark

People must realize that unless their entire family stays under house arrest for two years, we're all going to get this thing... right?

9:15 AM · Apr 21, 2020 · Twitter for iPhone

TEAM REALITY

Wes Pegden
@WesPegden

The recent antibody surveys in NY state show similar COVID prevalence among younger and older people. E.g., 13% in the age group 35-45 had antibodies, and also in the age group 75+. This raises a disturbing question: 1/7

5:39 PM · Apr 23, 2020 · Twitter Web App

Natural differences in transmission rates should lead to significant variation in infection rates with age, since younger people empirically have more possible transmission events per day, on average. Mitigations affect these contact patterns, however: ++ If mitigations more sharply curtail interactions among younger people than older people (by closing schools, bars, etc.) then the expected result is a shift in the distribution of infection to be more uniform among age groups, rather than shifted towards younger people... ++ This increases the IFR above what would be expected when younger people (due to their greater interaction rates) bear a greater burden from infection. IF a vaccine is around the corner, we shouldn't be worried by this as lot. HOWEVER, ++ IF this is a long haul, and NYC will eventually end up reaching something like herd immunity through an eventual lack of compliance with mitigations, or simply because compliance proves unsustainable over the long term, THEN this trend means that more people will have died, ++ compared with a strategy to reach herd immunity with a focus on affecting the age (and risk) distribution of the final immunity herd. These types of strategies are discussed in our preprint with @ChikinaLab: https://arxiv.org/abs/2004.04144 6/7 ++ Of course, these antibody surveys are still preliminary and we should interpret them with caution. It is possible to imagine relevant ways in which they could be biased.

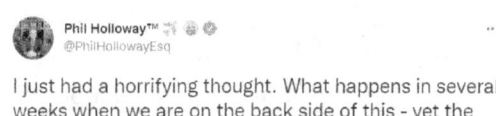
Phil Holloway™
@PhilHollowayEsq

I just had a horrifying thought. What happens in several weeks when we are on the back side of this - yet the virus is not eradicated and there is still no vaccine for #COVID19 ? Right now we are all thinking about this being over but when will it really be over?

7:57 AM · Apr 2, 2020 · Twitter for iPhone

Justin Hart
@justin_hart

Holy crap. This pic.

1:23 AM · Apr 19, 2020 · Twitter Web App

Martin Kulldorff
@MartinKulldorff

"Since COVID-19 operates in a highly age specific manner, mandated counter measures must also be age specific. If not, lives will be unnecessarily lost."
linkedin.com/pulse/covid-19...

	Scenario A: Equal probability of exposure		Scenario B: Younger are more exposed	
	RR	1/RR	RR	1/RR
9	0.0003	3560	0.0001	7120
'9	0.0034	297	0.0017	595
39	0.010	99	0.005	199
39	0.025	40	0.013	80
59	0.11	9	0.05	18
39	0.43	2.3	0.21	4.7
'9	1	1	1	1
	1.5	0.7	2.9	0.34

10:23 AM · Apr 10, 2020 · Twitter Web App

Stefan Baral
@sdbaral

I worry that #ShutItDown was an intervention designed by and for richer folks and actually means "shut it down unless it is something that I need and then just do that thing but shut it down except for that".

4:56 PM · Mar 22, 2020 · Twitter Web App

Prof Francois Balloux
@BallouxFrancois

The report by the 'Imperial College COVID-19 Response Team' has been released on March 16th (tinyurl.com/tcdy42y). Now that things have calmed down a bit, I felt I should produce a thread summarising my personal take on it. (1/15)

12:58 PM · Mar 17, 2020 · Twitter Web App

It was much anticipated as it is considered as the scientific rationale driving the UK government's strategy to mitigate the covid-19 pandemic. It was also allegedly shared with the White House. ++ The 'Imperial College COVID-19 Response Team' is possibly the best rapid pandemic response unit in the world. I was based in that unit between 2007-2012 and can testify that their scientists are technically superb. ++ The modelling is sophisticated. Though, the report is somewhat narrow in its remit and some conclusions may be 'best-case-scenarios', despite its conclusions being somewhat uncomfortable. The modelling is based on strong assumptions, some that could be more explicit. ++ Key assumptions include the fact that immunisation is long-term. This remains a big unknown it remains to be seen how long immunisation lasts for, following infection by SARS-COV-2. Some evidence suggests immunity might be relatively short-lived. ++ The report does not explicitly consider the economic/health impact of different mitigation measures. For instance, the scenario of a 18-month lockdown would devastate the economy and could hence reduce life expectancy beyond the toll that SARS-COV-2 might exert on its own. ++ The report concludes that the effectiveness on the Covid-19 pandemic of any one intervention in isolation is likely to be limited, requiring multiple interventions to be combined to have a substantial impact on transmission. ++ Two basic strategies are considered: (a) 'mitigation', which focuses on slowing but not necessarily stopping epidemic spread (i.e. 'Herd Immunity"), and (b) 'suppression', which aims to reduce case numbers to low levels and maintaining that situation indefinitely. ++ The report has been widely interpreted as being supportive of the 'suppression' strategy, which may entail a country-wide lockdown for up to 18 months. This may be a slight misunderstanding of a sophisticated piece, which acknowledges there is no easy solution. ++ Though, the report is in fact fairly balanced and acknowledges (but doesn't model) the extreme economic damage that a 'suppression' strategy would inflict on the economy (and by extrapolation to health, education and longevity). ++ The report can be read in many different ways, but fundamentally, it confirms that we are facing a series of uncomfortable options within a continuum ranging from paying a heavy death toll right now vs. an uncertain and possibly even worse future. ++ There may actually be no choice. Politically, but also morally and ethically, any option leading to a heavy death toll over the coming months, but offering the possible prospect of maximising long-term life-expectancy would be a difficult sell. ++ As such, I have little doubt we're heading for a 'suppression' approach to covid-19. Given the many unknowns and the difficult moral implications, it probably makes sense to try to address the most urgent issue. There are times when tactics might trump strategy. ++ Pandemics like this one have been looming for millennia, and covid-19 won't be the last humanity will face. We have been caught napping and we now have to deal from a position of weakness with a difficult situation that could have been, in principle, largely avoidable. ++ I predict some difficult times ahead in the immediate future. Though, I have some hopes that the COVID-19 pandemic will bring the best out of humanity, and that this crisis will act as a catalyst for us to deal more effectively, together, with future global challenges.

TEAM REALITY

Bethany S. Mandel ✓
@bethanyshondark

This is something that has been really bothering me for a while. In all of this, I don't think we've had anything resembling a conversation about how deeply unfair all of this is to kids. In an instant, we took everything away from them. School, friends, activities. All of it.

9:18 PM · Apr 13, 2020 · Twitter Web App

We have no answers for them about when it might come back. And on some things, like their activities, we can't even promise it will come back. Will their dance school survive this? The Little Gym? The indoor playspaces? The places they take school and family trips? ++ And my Facebook is filled with NOTHING but parents complaining about their kids. It's really self-centered behavior that we would never accept if it came from our own kids. ++ Kids aren't really getting sick, and yet we are asking them to sacrifice their childhoods because they might be carriers. As I tell my kids they can't go on a playground because someone might call the police on us or post our pictures on NextDoor to complain, I'm getting angry. ++ We are ripping everything away from them, complaining that they are reacting to it negatively, and we have no answers for them about any of it. ++ This is going to have a lasting impact on a lot of kids. Kids who needed special ed services who aren't getting them. Kids who live with abusive parents and who no longer have mandatory reporters looking out for them. Teenagers who were already having mental health struggles. ++ What I keep wondering is this: What's the end game here? So school is canceled the rest of the year and we're talking about "maybe" coming back in the fall. So summer camps are out (lets see if they survive a canceled season). Do we put their lives on hold until a vaccine? ++ You can't hit pause on a kid's development. You can't just say wait two years so we can develop a shot. It doesn't work that way. We are making a deeply unfair decision for an entire generation if that's the expectation. ++ At what point do we say that their lives have to be factored into this conversation? Into this calculus? Yes, reopening has its costs, that we have heard. But kids staying locked down like this has incalculable costs as well. We need to be discussing them.

David M
@ComradeDoom1

Replying to @HAgredano

Feels like we're all racing toward a cliff right now… an indefinite quarantine surely has the potential to provoke unprecedented civil unrest/riots in this country

11:07 PM · Apr 16, 2020 · Twitter Web App

Justin Hart
@justin_hart

Govt: "Those with co-morbidities are most susceptible to the virus. These ailments are triggered by low physical activity, lack of access to healthy foods, hypertension from stress, and… ohhhh… dang! Have you been inside this whole month, jobless, eating junk food!?
My bad."

2:55 AM · Apr 21, 2020 · Twitter for iPhone

Stefan Baral ✓
@sdbaral

Likelihood of asymptomatic outdoor transmission=negligible

Likelihood of adverse physical and mental health consequences of closing parks and green spaces= significant

Any chance for some specificity and local dynamics to inform public health decision making during #COVID19?

10:12 PM · Apr 6, 2020 · Twitter Web App

Mark Changizi
@MarkChangizi

"Of the 705 passengers who tested positive for the virus on the Diamond Princess, six died -- which is a death rate of less than 1%. All of the patients who died onboard were more than 70 years old."

Seems inevitable that the death rate is overestimated

11:54 AM · Mar 10, 2020 · Twitter Web App

TEAM REALITY

Prof François Balloux
@BallouxFrancois

There have been considerable discussions on the epidemiology of Covid-19, and some on its economic impact. Though, I suspect the possible political and ethical implications may not have been fully appreciated yet. In this thread I will allude on some potential issues. (1/8)

12:17 PM · Mar 20, 2020 · Twitter Web App

To get this out of the way, I'm not an ethicist nor a political scientist. Thus, please don't ask me for an expert opinion. I'm expressing myself as a simple citizen and consider that everyone should have an equal voice in the debate that may ensue. ++ I believe that even in the darkest hours, we need to protect our rights and values. We're entering unchartered territories, legally, morally and ethically. The world will arguably change, and it is our right and duty to contribute to shaping the world that will emerge. ++ It may seem sensible for governments to seize additional powers during a state of emergency such as the Covid-19 pandemic. Though, the pandemic may last for while, and we should not accept our rights to be eroded indefinitely. ++ Testing people for a 'live infection' is a diagnostic procedure and may be considered as ethically unproblematic in any circumstance. Testing for past Covid-19 infection (serology) poses interesting and to the best of my knowledge unprecedented moral and ethical challenges. ++ We might for instance imagine a public health policy where only people who have been exposed to SARS-cov-2 would be allowed to resume a normal life and go back to work. This could create a somewhat perverse incentive for people actively trying to get infected. ++ Allowing only the 'previously infected' to resume a normal life could pose privacy issues as it would in effect be akin to using medical information for discrimination. The irony is that it may lead to a situation were the 'uninfected' are those being discriminated against. ++ I could understand if there were not much thirst for a debate on ethics/morals as we're bracing ourselves to face the first Covid-19 wave. Though, it might be a mistake not to reflect on the wider societal aspects, if we hoped to shape a better world after the pandemic.

Wes Pegden
@WesPegden

Read our new piece: "A call to honesty in pandemic modeling", with @ChikinaLab:

medium.com/@wpegden/a-cal...

The basic message? Hiding infections in the future is not the same as avoiding them.

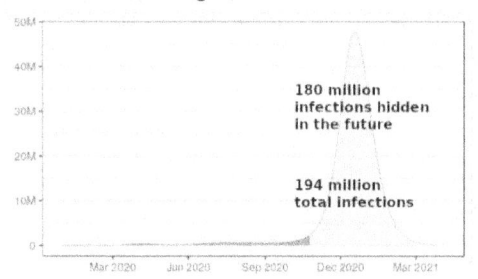

180 million infections hidden in the future

194 million total infections

11:09 PM · Mar 29, 2020 · Twitter Web App

Kyle Lamb
@kylamb8

I took every country and cross referenced the cases/deaths with ave. temp. I broke larger countries into regions using the biggest city for each, to account for differing climates. I found 58% of cases are occurring in 7-12 C with 87% of deaths in that range.

Temp F	Temp C	Cases	Deaths	Population	Cases per 1 mil pop	Persons per death
<32	0	5235	28	375,529,648	13.9	13,411,773
33 - 36	1 - 2	8021	92	209,912,981	38.2	2,281,663
37 - 39	3 - 4	35758	390	572,581,982	62.5	1,468,159
40 - 43	5 - 6	73531	996	413,404,870	177.9	415,065
44 - 46	7 - 8	117038	5110	489,915,905	238.9	95,874
47 - 50	9 - 10	75197	7858	283,071,330	265.6	36,023
51 - 54	11 - 12	47617	2866	197,135,311	241.5	68,784
55 - 57	13 - 14	15160	449	210,693,079	72.0	469,250
58 - 61	15 - 16	8719	148	444,572,381	19.6	3,003,867
62 - 64	17 - 18	5255	114	446,606,779	11.8	3,917,603
65 - 68	19 - 20	3152	43	195,636,616	16.1	4,549,689
69 - 72	21 - 22	5703	24	421,799,546	13.5	17,574,981
73 - 75	23 - 24	2387	19	424,775,248	5.6	22,356,592
76 - 79	25 - 26	2821	47	537,032,013	5.3	11,426,213
80 - 82	27 - 28	4163	83	966,776,946	4.3	11,635,867
>83	29+	1882	61	873,171,894	2.2	14,314,293

2:24 PM · Mar 25, 2020 · Twitter Web App

TEAM REALITY

John Ziegler @Zigmanfreud

A theme I'm seeing in conversations with friends who support @realDonaldTrump is growing frustration with his complete lack of consistent/coherent message & his total loss of balls. Liberals are dominating the messaging because their side is in lock step, but his side is confused

5:12 PM · Apr 26, 2020 · Twitter for iPhone

AJ Kay @AJKayWriter

Whether or not extending lockdowns would curb the spread of COVID isn't the right question.

The right question is, "Can we afford it?"

And the answer is, "No."

We can't afford the famine, unemployment, neglect of non-COVID healthcare, supply chain disruptions, economic ruin..

3:38 PM · Apr 25, 2020 · Twitter Web App

Phil Kerpen @kerpen

It's kind of weird how the same people who insist on not using drugs until they are proven safe and effective adamantly support non-pharmaceutical interventions like school closures that have no proven effectiveness and enormous known costs.

2:23 PM · Apr 17, 2020 · Twitter Web App

Bethany S. Mandel @bethanyshondark

I don't know who needs to hear this, but you don't need a mask in your own car or backyard.

2:39 PM · Apr 17, 2020 · Twitter Web App

Show Me The Data @txsalth2o

Replying to @RantyAmyCurtis and @KenShepherd

It's about domestic abuse
It's about hunger
It's about depression
It's about alcoholism
It's about opioids
It's not about a haircut.

2:17 PM · Apr 28, 2020 · Twitter Web App

John Ziegler @Zigmanfreud

Stats the media should add to virus data:

-Dangerous inmates directly released
-Increase in suicide attempts
-Impact of delay on medical treatments for non-coronavirus patients
-Increase in domestic abuse
-Increase in robberies
-Days of school/graduations lost
-Businesses lost

6:17 PM · Apr 18, 2020 · Twitter for iPhone

AJ Kay @AJKayWriter

We risk our lives by living.

Living is worth the risk.

Locking ourselves inside of our homes isn't living.

2:06 PM · Apr 16, 2020 · Twitter for iPhone

Phil Kerpen @kerpen

This paper finds nursing home residents comprise 57% of all COVID-19 deaths in Spain, 53% in Italy, and 45% in France.

itccovid.org/wp-content/upl...

Date	Source	Deaths of care home residents linked to COVID-19	% of total COVID deaths
10/04/2020	Official data	1,405	42
11/04/2020	Official data	6,177	45
11/04/2020	Official data	82	45
11/04/2020	Official data	156	54
6/04/2020	Survey by official institute (extrapolation)	9,509	53
8/04/2020	Media	9,756	57

8:15 AM · Apr 16, 2020 · Twitter Web App

Mark Changizi @MarkChangizi

Lockdowns were NOT common sense measures. They were hysterical reactions out of fear.

Here are 15 of just some of the reasons why it was not common sense.

(I'm not including all the reasons we have NOW to see they were a bad idea.)

11:38 AM · Apr 27, 2020 · Twitter Web App

Vinay Prasad, MD MPH @VPrasadMDMPH

When two people disagree about the evidence for a medical practice it is painful to listen to one side argue "the other just doesn't care about saving lives"

That is merely re-stating the disagreement. Both sides care, they just differ on what the evidence says works.

10:48 PM · Apr 30, 2020 · Twitter for Android

TEAM REALITY

Vinay Prasad, MD MPH
@VPrasadMDMPH

What I have learned from COVID19:

[Thread]

2:13 PM · Apr 22, 2020 · Twitter Web App

1. Public health is undervalued, underfunded, neglected, mistreated, and it is easy to get away with that for years, decades, but someday that ends up biting you in the ass ++ 2. CDC recommendations/ guidance to restrict early testing was a mistake. Being able to scale up testing remains a total disaster; Without testing, fundamental Qs remain unknown, Sad and honestly embarrassing for such a great nation ++ 3. Doctors love to give unproven, unpromising therapies based on lousy, awful data. No number of medical reversals and no history book will ever convince many that the harms of just giving things can easily outweigh benefits. Sad! ++ 4. When we do get around to running trials, we do a lousy job: No controls No blinding Awful, bias susceptible endpoints Too large a sample Too small a sample Duplicative and redundant trials Trials not proportionate to scientific promise Total disaster ++ 5. Docs love to hype early, uncontrolled, retrospective, missing data, busted time zero, confounding by indication garbage. We also love to say things like "appears promising"; "a clue"; "may"; "might" "could" ... to absolve us of our faith in these crap papers ++ 6. We are quick to imagine a new disease changes the entire playbook. Forget everything you know about the vent, blood thinners We don't realize that such claims are HIGHLY provocative and requires a lot of data ++ The burden of proof is on those who believe THIS IS DIFFERENT to prove a different strategy is superior. Instead, our experts just change institutional guidelines willy-nilly ++ 7. A few loud mouths can distort and distract all media coverage. Some folks have no shame in saying BS on TV. ++ 8. Even many academics prove to be 'inauthentic' and 'opportunists' tweeting about things they are inexpert in with false confidence. Craving the same, sad TV fame. Basically reading the newspaper one day and tweeting about it the next. Changing their twitter bios. lol. ++ 9. The worse the data supporting the policy, the louder some shout "The science says..." "We must..." "90 studies show..." Creating a hostile atmosphere for those who disagree. ++ BTW, citing the number of weak, tangential, irrelevant studies that support a claim is meaningless. Go back and look at how many studies support some now debunked claims. It can exceed 5k! ++ 10. If there are serious academics who disagree. My philosophy is hear them out, consider what they say, and then feel free to rule against them; instead, as @drjohnm and others have observed, I observe there is too much pre-occupation with silencing them. ++ Asking what their real motivation(s) is/are. Is it not possible they just disagree? Do we ask ourselves our motivations simply because we have reached a different conclusion? ++ 11. Preprints. I am not convinced that pre-prints are worse than usual misleading science, but boy do some try to ruin it for their kind ++ 12. There is no private, personal tragedy that someone else cannot make about themselves in an obnoxious, narcissistic tweet. ++ 13. Folks are lying about how good telework and tele-lectures are ++ 14. It is easy to minimize or discount economic damage when you are financially well off with a stable job; and easy to demonize folks who ask us to consider that economic damage actually does crush human health and well being

++ 15. HIPAA is a suggestion ++ 16. When you do something generous, noble or honorable, make sure to let others know about it on twitter ++ 17. Fear-mongering and false certainty drives social media traffic Nuance can #deleteyouraccount ++ 18. Explaining the need for RCTs to well educated physicians is painful "Would you be randomized" yes, of course "Would you want your loved one randomized" Yes "What if it is dire" All the more "But it can't hurt" ++ 19. When you get people to make important sacrifices, there is always someone willing to come along and ask for more sacrifices that aren't ++ 20. Some models were wrong. I am sorry to say. They were. They made a prediction based on what actually occurred and observed outcomes are outside all uncertainty or confidence intervals. They are not 'wrong' b/c things worked as intended. They failed to predict outcomes ++ 21. There is always someone who will violate professional and ethical norms ++ 22. Drug companies won't run the right trials until we make them ++ 23. It's important to calculate 95% CI correctly ++ 24. The sensitivity and specificity of the blue check mark is .3 and .5, respectively.

Show Me The Data
@txsalth2o

Replying to @AlexBerenson and @nytimes

94 Hospitals have announced furloughs. Ninety Four. Including MAYO.

Becker's is keeping a running account of them here.

beckershospitalreview.com/finance/49-hos…

6:58 AM · Apr 14, 2020 · Twitter for iPhone

Bethany S. Mandel
@bethanyshondark

Imagine the Mayor doing this about any other community in the city. You wonder why NYC has had an anti-semitism problem? Look to the top. This is dangerous rhetoric that could get someone killed. Also, Bill, STOP GOING TO BROOKLYN.

> **Mayor Eric Adams** @NYCMayor · Apr 28, 2020
> My message to the Jewish community, and all communities, is this simple: the time for warnings has passed. I have instructed the NYPD to proceed immediately to summons or even arrest those who gather in large groups. This is about stopping this disease and saving lives. Period.

10:38 PM · Apr 28, 2020 · Twitter for iPhone

Prof Francois Balloux
@BallouxFrancois

Unless we eradicated #COVID19 globally, it will stay with us. If you had a plan for global eradication, applicable to places such as India, Syria or Afghanistan, please let the world know. Otherwise, I suggest we move on to an adult discussion on how best to minimise harm.

3:18 PM · Apr 12, 2020 · Twitter Web App

Kyle Lamb
@kylamb8

I finally was able to run the temperature data for the 3,142 U.S. counties & equivalents against cases per capita. Population is 2018 U.S. Census estimate. Climate data is an average of each county for March 2020 (source: NOAA climate database). Five buckets of 628 counties each.

Number of COVID-19 cases of 3,142 U.S. counties & equivalent per 1 million
Cases current through Apr. 18
(Average temperature taken from NOAA for March 2020)

Range:	Average	Population	Cases	Cases per 1M
-4.4 F to 38.3	32.9	29.0 m	21,396	737
38.3 F to 44.8	41.6	79.7 m	238,506	2,991
44.8 F to 51.6	48.4	79.6 m	297,218	3,733
51.6 F to 59.4	55.2	62.8 m	44,924	716
59.4 to 76.4	65.5	74.6 m	89,774	1,202

9:07 PM · Apr 21, 2020 · Twitter Web App

Bethany S. Mandel
@bethanyshondark

Hello to every editor and producer who follows me: What do you have planned to cover the Mayor of NYC singling out the "Jewish community" in a tweet last night? Lashing out at the Jews in the middle of a pandemic has some historical precedent. Might want to look that up.

7:35 AM · Apr 29, 2020 · Twitter for iPhone

Bethany S. Mandel
@bethanyshondark

What would you have planned if the Mayor did the same about a pickup basketball game with some black and Hispanic residents? If he then with broad strokes condemned every black and Hispanic New Yorker? If he threatened them with the NYPD? Whatever you would have done, do that.

7:44 AM · Apr 29, 2020 · Twitter for iPhone

TEAM REALITY

Ann Bauer
@annbauerwriter

There's a virus but don't wear a mask b/c it won't help. It might even make you sick. DON'T WEAR A MASK. OK, wear a mask for peace of mind. You must wear a mask. Here's a mask you should make out of a bandanna. If you don't wear a mask, police will drag you off the bus.

9:44 PM · Apr 14, 2020 · Twitter Web App

2 million people will die in America. If we flatten the curve, maybe 1.3. FLATTEN THE CURVE! If you don't flatten the curve 1 million people will die. Or maybe 300,000. OK, 60,000. But if you stay home and kill the economy? Well, we're not making any promises. ++ This virus is deadly to older people and those with pre-existing conditions. We must keep those people safe! They should stay home. No, school children should stay home b/c they're vectors. (You know how germy children are) ++ Besides, younger people are dying. In droves! Everyone, stay home!!! You're killing all your grandmas and you're probably going to die too. Here's a 35 yo who just got off a ventilator and he is proof that you, Millennial, are going to die. Because you went on spring break. ++ Ooops. That was Gen Z. Our bad. But still! Millennials are the problem. You're out there, going to hot yoga and slobbering all over your craft cocktails and KILLING OFF THE GREATEST GENERATION! Fuck you. Go home. ++ So as we said, this thing is ravaging nursing homes and the average age of death is 79. You're stuck inside, unemployed, deprived of sunlight and eating Cap'n Crunch while watching Tiger King? Don't worry! You're young & healthy so you have a long life to get back on your feet. ++ Thing is, we're going to give everyone small business loans and that's going to keep them going 'til after this whole thing is over. I mean, we're going to give SOME people small business loans. ++ Except, jeez, sorry, we let some frat brother of Jared's write the code for the site and it's not working. At all. Also, sadly, we ran out of money in minute 47, but you have to agree it was a magnificent effort and a really nice thought. ++ But wait! Ventilators!!!! (squirrel!) We're making 'em out of spare parts. Did you see that scene in Apollo 13 where they made a filter out of dribs and drabs and saved the astronauts who were going to suffocate? Like that. Millions of ventilators. They're going to save us all. ++ Governor Cuomo alone needs 160,000. HE NEEDS THEM! People will die without them. Why can't he get them? OK, he needs 80,000. Well, maybe 40,000. OK, 20,000 but that's a big number too! ++ Only, wait, just in: Doctors are overusing ventilators. They're doing damage. Who knew? Patients under heavy sedation having their lungs artificially pumped to bursting? Mmmm. Not so good after all. It saves them! Some of them. About 18% of them. But there was that 35 year old... ++ But one thing we know FOR SURE is that this virus came from a 'wet' market in China. (Why is 'wet' in quotes? I have no idea! But it is - always.) This was a natural zoonotic transmission. It was one guy who ate a bat. Or maybe a pangolin. No government was involved. ++ It was natural, people! There's no X-Files shit going on here. Well, yeah, there was that one lab in Wuhan that was developing bat coronaviruses. But it's a coincidence! Pure & simple. Why don't you people understand? Just listen to us. Stay home. Cover your face. It'll be fine.

TEAM REALITY

Clifton Duncan: Good Looking Loser.
@cliftonaduncan

I'm not sure I'm on board with the emerging idea that "things will never be the same" after the #covid #crisis, and we should "just accept" it.

I understand the viewpoint, but it seems fatalistic to me; a mindset vulnerable to exploitation by powerful, unscrupulous actors.

12:00 PM · Apr 18, 2020 · Twitter for Android

Should we "just accept" government overreach, or greater state surveillance, or the possibility of an expanded police state? I've observed Progressives through to Conservatives concerned about these issues; can we not all agree as Americans that these things are bad? #lockdown ++ Nobody wants sickness or death; but could too much "safety" drive us to a less free society? I notice many of the same people agitating for an extended #lockdown mocked Republicans in the wake of 9/11 for their obsession with "keeping our country safe". ++ The point isn't right/wrong; but I'm seeing VERY LITTLE serious and open conversations about TRADE-OFFS, about COSTS vs. BENEFITS going forward in the wake of #Covid_19. It's either "you want people to die!" or "we're heading into an authoritarian nightmare!" Seriously??

Phil Holloway™
@PhilHollowayEsq

When I see people wearing masks all alone - outside - where there is ZERO threat of infection from #COVID19 or really anything else, I wonder things. Is it possible the emperor has no clothes?

11:23 AM · Apr 29, 2020 · Twitter Web App

AJ Kay
@AJKayWriter

Someone out there right now is saying, "Damn. That was easy."

SMDH.

2:40 PM · Apr 2, 2020 · Twitter Web App

AJ Kay
@AJKayWriter

I keep hearing, "We should listen to the experts."

As though there's only one field of expertise, two experts, they're always right, and they always agree with each other.

10:51 AM · Apr 16, 2020 · Twitter for iPhone

Bethany S. Mandel
@bethanyshondark

Your four-year-old doesn't need worksheets and you can't use an app to teach your child to read. Have parents always been this divorced from common sense education best practices? These are educated folks in the middle class.

8:18 AM · Apr 17, 2020 · Twitter Web App

TEAM REALITY

MAY 2020

John Ziegler ✓
@Zigmanfreud

As really horrible as things currently seem in America, they are actually going to get worse.

Maybe much worse.

Happy Friday!

10:50 AM · May 29, 2020 · Twitter for iPhone

Bethany S. Mandel @bethanyshondark

Remember when we were told we had to flatten the curve and we'd lockdown for a few weeks to ramp up PPE and free up ventilators or else we'd have to start death panels? When did that turn into indefinite lockdowns and economic destruction because "if it saves one life?"

11:58 AM · May 6, 2020 · Twitter for iPhone

This isn't about greed. It's survival. People can't buy food or pay rent or mortgages. Small businesses are closing. Dentists and doctors are going into the red. Schools are going to start closing. This is the destruction of society we're talking about. ++ There will be no pediatricians or general doctors or physical therapists or nurses or home health aides. No dentists. No zoos or aquariums. No private schools. No restaurants or caterers. No hairdressers or nail technicians. No gyms. No summer camps or daycares. ++ We never had ventilator shortages. My local pediatric ER converted to a COVID ward and now sits empty. What are we waiting on here? I'd genuinely like an answer. A vaccine? Because if that's it, our society will be absolutely wrecked in the meantime. ++ You can call me a Grandma killer. I'm not sacrificing my home, food on the table, all of our docs and dentists, every form of pleasure (museums, zoos, restaurants), all my kids' teachers in order to make other people comfortable. If you want to stay locked down, do. I'm not. ++ Doesn't mean it won't be done responsibly but I am just. I am done. I feel lied to about the terms of this lockdown and I regret ever trusting that it would be done responsibly.

Hold2 @Hold2LLC

While researching LTC COVID deaths to determine impact nationwide in regards to policy decisions and protective measures for the regular populace, I came across this major anomaly in NY. Look at this screenshot from this article: freopp.org/the-covid-19-n... @EthicalSkeptic 1/5

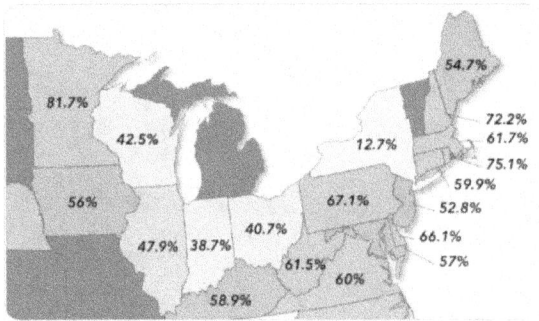

5:09 PM · May 27, 2020 · Twitter Web App

Apparently, NY stopped counting LTC deaths that occurred outside the LTC facility. In other words if an LTC resident contracted COVID but then died in a hospital, that is _not_ counted as an LTC death but rather a regular death. ++ In the original image, you can see this is a major outlier. "...of the nine states with the largest coronavirus outbreaks, New York has the lowest reported share of state deaths from long-term care residents at 20%, according to The Times." ++ This outlier skews the national average of LTC deaths down from 52.5% to 42+% (ref: http://freopp.org). If tracked properly, NY may _increase_ the national average considering % for surrounding states. 15,499 / 23,643 deaths in NY are 70+ (65.5%) ++ Chart showing US deaths divided between LTC and non-LTC. This is based on ratio, not based on dailies. I don't have daily LTC deaths, but that would be extremely valuable to see charted over time. How should policy and public discourse differ between these 2? @EthicalSkeptic

TEAM REALITY

Phil Kerpen @kerpen

Pennsylvania has more COVID deaths over age 100 than under age 45.

More deaths over age 95 than under age 60.

More deaths over 85 than under 80.

health.pa.gov/topics/Documen...

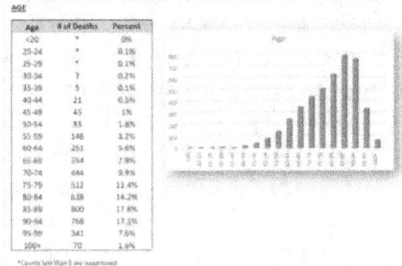

12:34 PM · May 19, 2020 · Twitter Web App

Gummi Bear @gummibear737

This is potentially huge. They've found that 40-60% of uninfected people could have partial immunity to Covid due to previous infections with other "common cold" coronaviruses.

This is a good thread to get the basics.

Full study here: cell.com/cell/fulltext/...

12:48 PM · May 26, 2020 · Twitter for iPad

Josh Stevenson @ifihadastick

Testing Rate vs Incident Rate in the US. The more we test the more we find. Initially testing was about understanding the spread. Now, what? Are we in an endless cycle of confirmation bias driven by the obsession on "total cases"? Is this a contest now?

@FatEmperor

9:42 PM · May 27, 2020 · Twitter Web App

Virál Myãlgía MD, PhD @contrarian4data

Replying to @BuffyWicks and @LorenaSGonzalez

There is no statistical correlation between date of lockdown or extent of lockdown and deaths per million. It was understandable not to know that from the outset. It's concerning to continue this assertion. Unfortunately, the data were in more than a month ago.

2:12 PM · May 10, 2020 · Twitter for iPhone

Phil Holloway™ @PhilHollowayEsq

Fear is very powerful. Sometimes it's misplaced- sometimes not. Whether current levels are justified is for history to decide. However I'm certain the current level of fear over #COVID19 was heightened by doom-predicting press conferences for the last several weeks.

6:48 PM · May 2, 2020 · Twitter for iPhone

Abir Ballan @abirballan · May 31, 2020
@SirKenRobinson

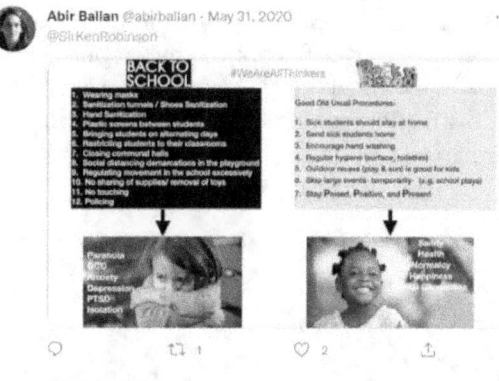

Mark Changizi @MarkChangizi

Some have argued that Sweden pretty much locked down organically, & so it's not much different than lockdowns.

First: CHOOSING to stay home and being FORCED to stay home are NOT EQUIVALENT, FFS!

Ahem, and...

Second: Google mobility shows that Swedes carried on mostly as usual.

10:53 AM · May 23, 2020 · Twitter Web App

TEAM REALITY

Prof Francois Balloux @BallouxFrancois

The lack of a resurgence in #COVID19 cases following the easing of lockdowns in several countries is intriguing. I'll take Switzerland as an example. The lockdown ended on May 11 (schools/restaurants opened). Yet this did not translate in any increase in new cases so far.
(1/6)

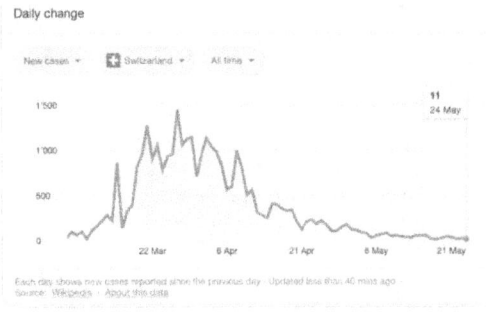

4:39 AM · May 27, 2020 · Twitter Web App

There are some residual 'social distancing' measures in place, which are not particularly strongly enforced and life is essentially back to the 'old normal. So why did the epidemic not resume? I can think of three non-mutually exclusive speculators reasons. ++ 1. 'Residual social distancing': People may still be behaving in ways limiting viral transmission. This feels unlikely as the main explanation for Switzerland at least, where life is essentially back to normal and hardly anyone wears a mask. ++ 2. Seasonality: The other endemic coronaviruses causing 'common colds' are highly seasonal with a marked peak in winter. Thus, if #SARSCoV2 transmission were similarly seasonal the #COVID19 epidemic may be slowed by the current summer weather. ++ 3. 'Cross-immunisation': A proportion of the population might have pre-existing immunity to #SARSCoV2, potentially due to prior exposure to 'common cold' coronaviruses. Under such a scenario, only a fraction of the population could be infected by the virus at this stage. ++ 'Residual social distancing' and 'seasonality' both likely contributed to the lack of resurgence of #COVID19 after the easing of lockdowns. 'Cross-immunisation' is a 'wilder' hypothesis but might explain some intriguing observations (e.g. no transmission between spouses).

Stefan Baral @sdbaral

Me 5 months ago: Future is virtual teaching at all levels, virtual conferences, and virtual working group meetings.

Me now: I hate virtual teaching, what is the difference between a virtual conference and webinars, and virtual working meetings exclude the most important voices

9:33 AM · May 10, 2020 · Twitter Web App

Kyle Lamb @kylamb8

Someone didn't believe me when I told them that more stringent lockdown states actually are faring (much) worse. 49 states + DC grouped by days in stay at home order or non-essential business closures (N.Y. excluded for fairer comparisons). Check this out...

Covid-19 cases and deaths per capita		
Grouped by no. of days with stay at home order or business closures		
Group 1		
54-59 days, 55 average		
Includes: CA, CT, DE, IL, LA, MI, NJ, NM, OH, OR, PA		
492 cases / 100k	32 deaths / 100k	14.3% positive tests
Group 2		
46-53 days, 51 average		
Includes: CO, KY, MA, MN, NH, TN, VA, VT, WA, WI		
396 cases / 100k	20 deaths / 100k	11.4% positive tests
Group 3		
37-45 days, 41 average		
Includes: AZ, DC, HI, IN, MD, ME, MT, NC, NV, RI, WV		
313 cases / 100k	16 deaths / 100k	10.6% positive tests
Group 4		
20-37 days, 30 average		
Includes: AK, AL, FL, GA, ID, KS, MO, MS, OK, SC, TX		
210 cases / 100k	8 deaths / 100k	6.7% positive tests
Group 5		
No shutdown		
Includes: AR, IA, NE, ND, SD, UT, WY		
305 cases / 100k	5 deaths / 100k	8.2% positive tests

6:00 PM · May 16, 2020 · Twitter Web App

Justin Hart @justin_hart

A reminder that COVID-19 is NOT like the flu in one important way... if it were:
- 1000 kids would now be dead
- 200 infants would be deceased

As it is - we have fewer than 20 deaths across those brackets.

9:25 AM · May 27, 2020 · Twitter for iPhone

TEAM REALITY

Wes Pegden @WesPegden

A short thread on why I think it is important to engage seriously with analysis of herd immunity strategies regardless of whether you think we should be pursuing them right now. (Short version: understanding them may help prevent the most COVID deaths in the long term). 1/7

10:42 PM · May 4, 2020 · Twitter Web App

The first thing to emphasize is that not all herd immunity is the same. The pandemic flu of 1918—where death rates varied among U.S. cities by a factor of 3 depending on how they implemented mitigations, even though all reached herd immunity—is a historical demonstration. ++ For COVID in particular, the orders-of-magnitude variation in the mortality rate by age (anti-correlated with transmission rates) makes it plausible that we can have an order-of-magnitude effect on mortality and morbidity even within the context of reaching herd immunity. ++ But wait. Isn't it better to prevent all of those deaths, through containment measures like test+trace? Definitely! If we have high confidence they will work. But if there is a realistic chance containment will falter after e.g., a year, and transmission rates increase... ++ .. in chaotic ways, then we would risk reaching herd immunity "by accident", which, again, could be an order of magnitude worse than reaching it with targeted strategies. The current high compliance with mitigations is a precious resource for affecting the herd! ++ Actual responses will depend on a mix of ideas. Immunity phenomena give MORE promise to the utility of measures like test+trace: At first glance, Singapore's experience with test+trace has been disheartening, despite apparent structural advantages in climate, gov, geography... ++ But if we take seriously the idea that heterogeneity can make immunity matter at thresholds significantly lower than suggested by simplistic homogeneous models, it should increase our confidence that test+trace can be more effective in places with higher levels of immunity.

Kyle Lamb @kylamb8

From a May 21 editorial in the New England Journal of Medicine, "Universal Masking in Hospitals in the Covid-19 Era."

This is consistent with Dr. Fauci commenting on March 8, "masks might make you feel better" but they don't offer you the protection you think they do.

a mask outside health care facilities offers little health authorities define a significant exposure et with a patient with symptomatic Covid-19 th d some say more than 10 minutes or even 30 mi a passing interaction in a public space is theref despread masking is a reflexive reaction to anxi

6:57 PM · May 26, 2020 · Twitter Web App

Bachman @ElonBachman

I don't think so many people have ever been so wrong about something so important, yet been so silent when confronted with overwhelming evidence of their error.

For three months we believed, against all evidence, that lockdowns worked.

And now, silence.

9:47 PM · May 28, 2020 · Twitter Web App

Bethany S. Mandel @bethanyshondark

43% Of COVID-19 Deaths Are In Nursing Homes & Assisted Living Facilities Housing 0.6% Of U.S.
forbes.com/sites/theapoth...

6:43 AM · May 27, 2020 · Twitter for iPhone

Clifton Duncan: Good Looking Loser. @cliftonaduncan

Isn't it interesting that the people who smugly pledge fidelity to "facts, science & data" regarding lockdowns are the same people who absolutely ignore the "facts, science & data" regarding the wide-ranging and long-term devastation caused by economic collapse?

8:02 PM · May 12, 2020 from Manhattan, NY · Twitter for Android

AJ Kay
@AJKayWriter

If you want to "listen to the experts", listen to me.

I'm an expert in narrative & there is 100% a media narrative re: COVID.

Narrative is storytelling & it's different than journalism. Journalism informs, sans judgement.

Pull up latest headlines. Look for the adverbs..1/5

11:47 AM · May 7, 2020 · Twitter for iPhone

... Look for the emotional triggers: the pleas to fear & virtue. Narratives use those elements to engage the reader. Reporting increase in cases w/o ref to the significance of the concurrent increase in testing is an example of an engaging narrative, as is the singular.. ++ focus on COVID as the only threat to our way of life. It prioritizes story arc over facts...follows a plotline. Disguising narrative as journalism is deceptive. It's akin to native advertising—in which I'm also an expert, not b/c I do it professionally, but b/c I won't.. ++ ...b/c it erodes brand trust, minsinforms, & steals the earned attention of consumers. This narrative has a similar deceptive angle but is far more harmful. It blinds ppl to the whole of the risks we are facing & weaponizes self-righteousness to silence dissenting ideas.. ++ Figuring out "why" the media is crafting the narrative is not in my wheelhouse. That's up to someone else. But if you insist on listening to experts only, listen very closely to me: The media is pushing a narrative. ++ ETA: This is not an easy assertion for me to make. I have an penchant for comfortable naïveté & high evidentiary threshold. I have no political alignment, nor covert agenda. What I am after is the truth. And to admit that this is the truth is profoundly disquieting.

Mark Changizi
@MarkChangizi

The Lockdown religious cult.

~ Believed initially on faith ("common sense")
~ Impervious to evidence they did nothing
~ Demand that all else must be sacrificed
~ Requires unrelenting devotion and asceticism
~ Promises forever life
~ Moral outrage for any who protest

5:39 PM · May 24, 2020 · Twitter Web App

Erich Hartmann
@erichhartmann

We must re-open NOW! pscp.tv/w/cYvthDI0NjU2...

9:28 AM · May 14, 2020 · Periscope

Clifton Duncan: Good Looking Loser.
@cliftonaduncan

I truly don't understand how #COVID became a political issue.

I also don't understand why people can't see the utter stupidity in MAKING it a political issue.

Is the real disease partisanship?

Tribalism?

11:03 AM · May 11, 2020 · Twitter for Android

Justin Hart
@justin_hart

A reminder of HOW BAD the IHME/Murray model was.

Look - I've built models which were wrong. But we didn't presume to dictate billions of dollars of state budgets, health policies, & federal planning. (and fear-mongering)

Our scarce resources deserve better models!
>line = IHME

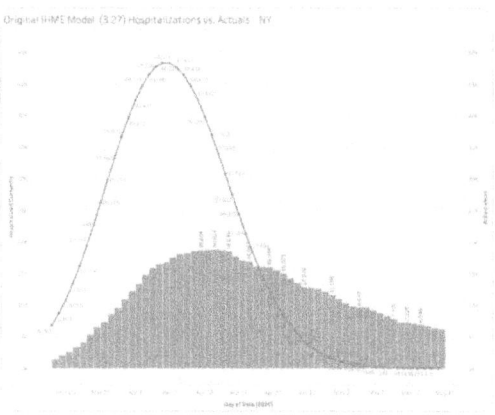

10:43 PM · May 19, 2020 · Twitter Web App

TEAM REALITY

Emma Woodhouse @EWoodhouse7

Hospitalization rate for school-age kids for C19 per @CDCgov. Open the schools @GovPritzker

> **Phil Kerpen** @kerpen · May 29, 2020
> That is 1.9 hospitalizations per 100,000 school age children.
> Show this thread

6:30 PM · May 29, 2020 · Twitter for iPhone

Bethany S. Mandel @bethanyshondark

We're setting national policy based on New York's numbers, where they hotboxed vulnerable seniors in nursing homes and just thought to start cleaning the main transmission vessels this week.

BuT wHaT aBoUt ThE bEaChEs In FlOrIdA

11:13 AM · May 6, 2020 · Twitter Web App

Clifton Duncan: Good Looking Loser. @cliftonaduncan

Dismiss #COVID19 as a hoax, you look dumb to any who understand the virus.

Dismiss reopening as selfish, you look dumb to any with a slight grasp of economics.

But oddly most who want to reopen acknowledge the virus, while most who want extended lockdowns dismiss the economics.

1:30 PM · May 22, 2020 · Buffer

Jennifer Sey @JenniferSey

This is criminal." The coronavirus pandemic is interrupting immunization against diseases including measles, polio and cholera that could put the lives of nearly 80 million children under the age of 1 at risk."

1:05 AM · May 23, 2020 from San Francisco, CA · Twitter for iPhone

Bethany S. Mandel @bethanyshondark

I'm starting to think if parents freaking out about this rare COVID complication had ANY idea how many kids die from the flu they'd never leave the house. If they knew how many die in car accidents they'd never drive. If they knew how many drown they'd never go near water.

5:57 PM · May 11, 2020 · Twitter for iPhone

Gummi Bear @gummibear737

This is the unreported tragedy of all the panic porn that's been circulating for months.

Undiagnosed/untreated cancer, heart disease, stroke and diabetes will cause more long term deaths than Covid.

9:13 AM · May 28, 2020 · Twitter Web App

Phil Kerpen @kerpen

Throw out the models and use the reams of actual observational data we have. Lockdowns have very limited effect. You can debate whether they increase or decrease spread slightly. We know the costs are MASSIVE. They is no fact-based case for continuing them.

2:29 PM · May 22, 2020 · Twitter Web App

John Ziegler @Zigmanfreud

Do you notice how the burden has dramatically shifted now?! "Flatten curve" to get hospitals ready has morphed into: "Unless you somehow prove to us that we should grant you your rights back, you can't have them."

Our rights just became privileges granted at whims of the state!

3:19 PM · May 12, 2020 · Twitter for iPhone

TEAM REALITY

JUNE 2020

Bethany S. Mandel ✓
@bethanyshondark

They killed tens of thousands of seniors in nursing homes. They won't let us go out of our own homes for months. They've destroyed the economy. And now if you don't do what they've been telling us not to do all this time - to gather with other people - we're irredeemable racists.

9:46 AM · Jun 4, 2020 · Twitter for iPhone

TEAM REALITY

Josh Stevenson @ifihadastick

Mostly white, affluent politicians, living their lives in extreme safety, free of oppression, launched into a lock-down knowing they could weather. Never thinking about those to whom "safety" is a dream. Whose mobility sustained their very lives.

@FatEmperor @EthicalSkeptic

11:34 AM · Jun 1, 2020 · Twitter Web App

Michael Tracey @mtracey

I listened to Gavin Newsom's full press conference yesterday in which he speculated all the various reasons for California's spike in COVID, such as family gatherings. Curiously, the mass protest movement -- reputed to be the most sweeping in 50+ years -- received no mention

9:32 AM · Jun 30, 2020 · Twitter Web App

Emma Woodhouse @EWoodhouse7

IL residents, hope we can agree @GovPritzker's criteria for moving the Phase 5 is illogical

✗ Vaccine is awhile off, if it ever comes
✗ "Readily-available treatment option" hoop gives no measurable parameters
✗ NO "new cases over a sustained period" doesn't even happen with flu

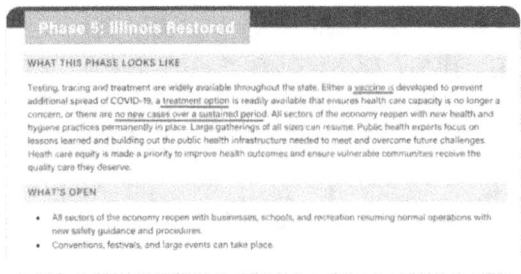

6:10 PM · Jun 29, 2020 · Twitter Web App

John Ziegler @Zigmanfreud

Just now, a police car drives near our neighborhood with sirens blaring...

Unprovoked and without skipping a beat, my 8-year daughter casually says to me...

"Someone must not be wearing a mask..."
#NewNormal

8:00 PM · Jun 15, 2020 · Twitter for iPhone

Stefan Baral @sdbaral

Sufficient experience with #influenza vaccine tells us that a #COVID19 vaccine will only reinforce disparities observed to date unless we start meaningfully addressing the underlying inequities driving these disparities. And that is a lot harder than drawing circles in a park.

8:50 PM · Jun 29, 2020 · Twitter Web App

Bachman @ElonBachman · Jun 29, 2020
Indeed. a WHO lit review found "No historical observations or scientific studies that support the confinement by quarantine..to slow the spread of influenza"

"This mitigation measure should be eliminated from serious consideration"

Yet we were told to "Follow the science"

John Ziegler @Zigmanfreud

I honestly didn't think that 2 weeks after massive/widespread protests/riots the news media would be able to pull off being able to BOTH overhype virus "spikes," AND not blame the protests/riots even a little bit.

However, even I underestimated their brazen lack of integrity...

12:36 PM · Jun 16, 2020 · Twitter for iPhone

Justin Hart @justin_hart

From my home county of Contra Costa. I kid you not.

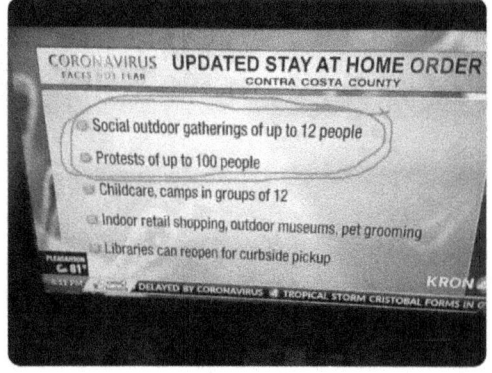

10:01 PM · Jun 5, 2020 · Twitter for iPhone

Vinay Prasad, MD MPH @VPrasadMDMPH

The French are running DISCOVERY
The UK RECOVERY and Remap-cap

And the US... A NIAID study subsidizing Gilead's R&D, so Gilead can run a trial of 5 vs 10 days, And docs treating thousands off label learning nothing daily.

US dominance in biomedicine is on trial in 2020

1:29 PM · Jun 16, 2020 · Twitter for Android

TEAM REALITY

Justin Hart
@justin_hart

A short rant:

Someone on Facebook challenged my continuous theme of "everyone settle down" with an inquiry: "well, when should we be worried? What stats will worry you?"

I feel strongly that this entire premise is faulty. I mean why are we still here talking about all this? 1/

8:19 AM · Jun 28, 2020 · Twitter for iPhone

The CDC just published their burden estimates for the 2019-2020 flu season. As many as 62K people died here in the US. Tha sort of thing has happened every year forever and we've never spoken of it except an occasional headline and a warning from grandma and the principal. ++ Probably 400+ kids died from the flu this season. That's 20x the number of kids who have died from #COVID19. No stories in NYT on those families. It's almost a given that suicides during the first month of the shutdown will have killed more high schoolers than C19 ever will. ++ When all is said and done the numbers of life years lost from influenza will far outweigh that from COVID. So I don't think we should be up late worrying about this. The virus is gonna do what it's going to do like every virus does. This is not the pandemic they sold us. ++ At this point this is a manufactured crisis that has seriously harmed the country for the long term. The sooner we can look away from the dashboards the better. As an addendum... I wrote this article on March 18th. Please, let's get back to life...

https://justin-hart.medium.com/the-coronavirus-dashboards-will-doom-us-all-7dbbaf419d6f

Kyle Lamb
@kylamb8

I'm amused at the people trying to act like doing your own work and research, mining, analyzing and citing data is actually a bad thing. It's no wonder this country has become so dumbed down. Try thinking for yourself and researching things. You may actually learn something new!

1:20 PM · Jun 20, 2020 · Twitter Web App

Bethany S. Mandel ✓
@bethanyshondark

Prediction: sometime in the future we'll see and maybe even admit how pointless the cloth masks are.

8:31 PM · Jun 18, 2020 · Twitter for iPhone

AJ Kay
@AJKayWriter

Per NYT, masks even one day/year (Halloween) are "disorienting" for kids.

"We're playing here with complicated neurological systems, with effects on the brain that combine nature and nurture, and, perhaps inevitably, with social and cultural overtones."

nytimes.com
How Children Learn to Recognize Faces (Published 2018)

12:37 PM · Jun 12, 2020 · Twitter Web App

Mark Changizi
@MarkChangizi

Q: Why did country X do well?
A: Lockdowns.

Q: Why did Y, who locked down, NOT do well?
A: Shoulda locked down harder.

Q: Why did Z, who didn't lock down, do well?
A: It locked down organically.

Q: Are there other possible answers?
A: Lock. . . .Down

2:54 PM · Jun 25, 2020 · Twitter Web App

Prof Francois Balloux ✓
@BallouxFrancois

The focus on science during the #COVID pandemic will damage the reputation of a number of established scientists. I can think of five mutually non-exclusive routes to scientific downfall.
1. Hubris
2. Rush to judgment
3. Dogmatism
4. Misconduct
5. Poor communication skills

4:07 PM · Jun 6, 2020 · Twitter Web App

TEAM REALITY

Wes Pegden
@WesPegden

The AAP released it's guidance for the re-opening of schools in the fall.

services.aap.org/en/pages/2019-...

"... the AAP strongly advocates that all policy considerations for the coming school year should start with a goal of having students physically present in school."

Thread. 1/10

7:07 PM · Jun 29, 2020 · Twitter Web App

Their document touches on several points, including the crucial role education plays in long-term outcomes, the uncharacteristically minor role children play in COVID transmission, the importance of planning and adaptability, and the impossibility of eliminating all risk. ++ Further, they discuss mitigating the risk posed by schools to the adults who work there, to whom the greatest risk may actually be posed by other adult staff. In all the COVID conversations happening now, the one about schools is not one that anybody should be sitting out. ++ Have you made it this far into the thread? Congratulations! You can read! Maybe you learned to read at home? Good for you! But that is atypical. 2/3 of children entering Kindergarten do not know the sounds letters make, and 1/3 cannot recognize the letters themselves. ++ In the U.S., literacy is a real problem with real consequences. Roughly 1 in 7 adults is estimated to have "below basic" literacy level by the NCES. This kind of literacy deficit translates into significant barriers to public health, with cascading consequences. ++ In all likelihood, we will be able to see the geographic landscape of the school-reopening politics of 2020 in the geography of adult illiteracy in 2050. PhD theses will be written on the topic. In the meantime, what is #epitwitter to do about it? ++ A lot of people who are good at thinking quantitatively about evidence have been sitting out the conversation on schools. This has meant that the conversation is driven largely by anecdotes and an ignorance of evidence regarding school transmission from around the world. ++ I think it's understandable to think that society generally has a tendency to not take a pandemic seriously enough, which can lead some commentators to feel out of place ever explaining that some things may be less risky than might naturally be expected. But this also harms. ++ Above all, I think that people who know better should not allow an irrational paranoia and fear to continue to grow among parents and teachers about the dangers of teaching children. It easy to to underestimate not only the damage one missed or curtailed school year... ++ will inflict, but also the difficulty we'll face EVER getting some students and teachers to return to the classroom, after a culture of fear around school is allowed to fester. Polls indicating more than a third of parents afraid of school should scare us all.

Angry Cardiologist
@AngryCardio

People need to accept the fact that reopening will be either too early or too late.

There is no way to "thread the needle". We don't even know where the needle is, let alone if there even is one.

9:01 PM · Jun 2, 2020 · Twitter for iPhone

Clifton Duncan: Good Looking Loser.
@cliftonaduncan

Replying to @NYGovCuomo

Wake up, New Yorkers:

Cuomo contributed to the highest concentration of #Covid deaths in the US with an order that needlessly put the elderly (the most vulnerable population) at risk.

His supporters are in denial.

11:54 AM · Jun 29, 2020 · Twitter for Android

TEAM REALITY

Prof
@covidtweets

All those who say lockdowns worked by showing that cases decreased after (they began decreasing before but that's another thread) are making a very basic scientific fallacy. (1/x)

6:07 PM · Jun 29, 2020 · Twitter for Android

Imagine you have a cold (or any other condition) against which your body's defense systems are perfectly effective. But it takes time before you fend it off. So you begin feeling bad and it gets worse. You cannot take it anymore so you go get a pill.

++ But because you felt the worst at the peak illness, if you simply could have waited a little more, you were about to begin feeling better. Now that you took the pill, and began feeling better, you conclude that the pill worked. ++ This is exactly what happened with the lockdowns. When it got really bad in many places, and hit the peak, they announced the lockdown. Some places did reactive lockdowns and didn't get hit hard, but they were never going to. ++ Common-sense measures were all we needed. Instead, we chose to destroy the society, because modelers, who likely knew some math and coding but nothing about coronaviruses, said so. ++ When the predicted doom never came, they applauded the "pill", despite predicting much worse outcomes even with the lockdowns. Nobody called them out because everyone was busy being scared to death that someone would infect them. ++ As @MLevitt_NP2013 pointed out, this was a total disgrace for science, and will erode the trust in science immensely for the years to come when the dust settles. People are just still too scared to realize.

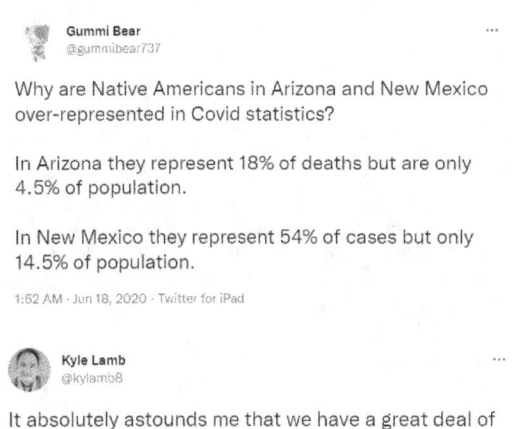

Kyle Lamb
@kylamb8

It absolutely astounds me that we have a great deal of data and know without a shadow of a doubt the 5-10% of our population that is at high risk, and we obsess over how we can micro manage the 90-95% who aren't. This behavior is a disease.

1:39 PM · Jun 29, 2020 · Twitter Web App

Prof Francois Balloux
@BallouxFrancois

Factoid of the day: The 'Russian flu' pandemic (1889/90), which killed around 1 million people at the time may have been caused by a coronavirus which is now one of the four 'common cold' coronaviruses (HCoV-OC43), causing winter epidemics year in, year out.

10:45 AM · Jun 5, 2020 · Twitter Web App

Hold2
@Hold2LLC

If anyone has requests for the charts, please reply here. Options include:

Easy:
- Single-state charts showing Case/Test/Positivity/CFR or any combination of data with no more than 2 different axes (scales)

Moderate:
- Single-state IFR-estimation chart

Hard:
- Multi-state IFR

11:26 PM · Jun 27, 2020 · Twitter Web App

Emma Woodhouse
@EWoodhouse7

Just got back from a large community playground with my daughter. I pulled off & picked up all the yellow caution tape & put it in the garbage, where it belongs.

Time to open up @GovPritzker @MayorHagerty

Stop punishing children

3:45 PM · Jun 15, 2020 · Twitter for iPhone

TEAM REALITY

Emma Woodhouse @EWoodhouse7

I'm confused about why @chicagosmayor felt last week's @ReOpen_Illinois protest posed a public health risk & needed to be dispersed, while this weekend's protests (& rioting) did not not. 🤔

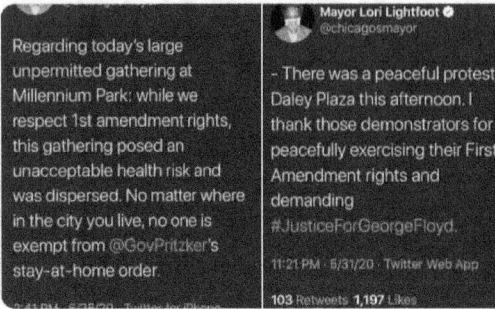

5:35 PM · Jun 1, 2020 · Twitter for iPhone

Emma Woodhouse @EWoodhouse7

Replying to @sebasT1026 @chicagosmayor and @ReOpen_Illinois

"My" cause? I didn't attend the @ReOpen_Illinois rally. I care deeply about justice in the murder of George Floyd. I also care about free speech, employment, & businesses being open. Fact is, LL was against protests, in the name of public health, until she agreed w/the cause

12:15 PM · Jun 2, 2020 · Twitter Web App

Clifton Duncan: Good Looking Loser. @cliftonaduncan

After this year, I never want to hear the phrases "science and data" or "listen to the experts" ever again.

6:34 PM · Jun 25, 2020 · Twitter for Android

Bethany S. Mandel @bethanyshondark

Watching people lose their businesses, bury family members who died alone without funerals, seeing special needs kids suffering without therapy ... while watching the same politicians cheering massive protests through the streets. The anger I feel will never, ever go away.

8:00 AM · Jun 15, 2020 · Twitter for iPhone

Gummi Bear @gummibear737

Instead of our politicians behaving as serious adults and basing policy on data, they continued and are still continuing to use lockdowns despite catastrophic consequences.

This is greatest political failure since the great depression

8:09 AM · Jun 23, 2020 · Twitter for iPad

Ann Bauer @annbauerwriter

I was the single mom of a son with autism & mental illness. I loved him wildly but life was brutally hard & we survived only thx to friends, family, teachers... If lockdowns had happened when my son was alive, we wouldn't have made it. I keep picturing pp like us, alone today

9:59 AM · Jun 24, 2020 · Twitter Web App

Phil Kerpen @kerpen

Hey @NYGovCuomo how about a sculpture of your nursing home deaths spiking 35x higher than the same period last year? (3x worse than Detroit.) Or heck we don't need a sculpture. Just give us an honest count of LTC deaths.
jamanetwork.com/journals/jama/...

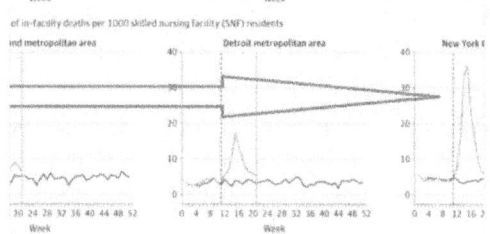

3:28 PM · Jun 29, 2020 · Twitter Web App

Gummi Bear @gummibear737

Sunetra Gupta, prof of theoretical epidemiology at Oxford University:

"We can't just think about those who are vulnerable to the disease. We have to think about those who are vulnerable to lockdown too. The costs of lockdown are too high at this point."

2:35 AM · Jun 6, 2020 · Twitter for iPad

Kyle Lamb @kylamb8

Among 15 to 24 year olds, less than 300 are hospitalized with Covid-19 for every million. Among every 100 hospitalized, 99.3 survive.

You have no idea how much I'm exhausted from the sports media acting like a 18-22 year old healthy athlete is getting a death sentence with this.

7:10 PM · Jun 19, 2020 · Twitter for Android

Bethany S. Mandel @bethanyshondark

I've buried a lot of people. Parents, grandparents etc. The pain is unimaginable. I cannot wrap my mind around it being made worse. Not being allowed to say goodbye. Not allowed to have a funeral. And for what? To see these images happening simultaneously.

8:10 AM · Jun 15, 2020 · Twitter for iPhone

JULY 2020

Prof Francois Balloux ✓
@BallouxFrancois

Beyond the loss of life and economic consequences of #COVID19, I am concerned about the level of fear in the population. I don't believe fear is ever healthy but I worry more about what will happen when it recedes. Fear doesn't just fade away, it morphs into resentment and anger.

10:37 AM · Jul 10, 2020 · Twitter Web App

TEAM REALITY

Hold2 @Hold2LLC · Jul 29, 2020
US Update: 7/29/2020

Graph 1: Daily Deaths + Current Hospitalizations

- CTP adjusted to TX/CA Hospitalization reporting from the CDC->HHS change, I am not sure how much of the drop is real
- Deaths up 83 W-o-W
- Expect deaths to increase still but Cases/Hosps have peaked

/1

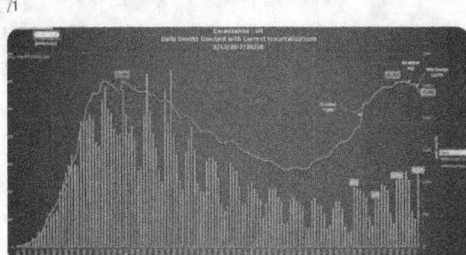

Jennifer Cabrera @jhaskinscabrera

Our society is so broken that we are being forced to do something that has no supporting evidence (masks) and forbidden to do something that has some supporting evidence (HCQ).

And it's all being driven by people who know nothing about how scientific inquiry works.

11:22 PM · Jul 29, 2020 · Twitter for iPhone

Real Developments @pdubdev

30-40% of urban kids will dropout of school due to distance learning.

LA Unified reported 40,000 high schoolers never attended one online session.

Not even one.

8:30 PM · Jul 9, 2020 · Twitter for iPhone

Prof @covidtweets

There is a surge in the South.

CNN: The irresponsible governors gambled with people's well-being, put money before saving lives, and recklessly reopened early.

There is a surge in the Asia-Pacific region.

CNN: We cannot explain it.

> **Prof** @covidtweets · Jul 29, 2020
> Virus is gonna virus:
> "Across the Asia-Pacific region, where countries were among the first hit by the virus and the first to contain it, there have been new and in some cases seemingly unexplained increases in the number of infections."
> cnn.com/2020/07/28/asi...

11:22 AM · Jul 29, 2020 · Twitter for iPhone

Ian Miller @ianmSC · Jul 29, 2020
iF wE aLl WeAr MaSkS wE cAn BeAt CoVID

> HEALTH AND SCIENCE
> **Hong Kong is 'on the verge' of an outbreak that could lead to hospital system's collapse, chief executive says**
> PUBLISHED WED, JUL 29 2020·1:53 AM EDT
> UPDATED WED, JUL 29 2020·3:18 AM EDT

Stefan Baral @sdbaral

Can we move past distractions of these parties and start talking about the deep inequities driving #COVID

Ie, shaming and criminalizing people is not helpful. Understanding why people are at risk and then addressing that risk is.
theverge.com/21324034/covid...
@zeynep @JuliaLMarcus

> theverge.com
> "COVID parties" are a pandemic urban legend that won't go...
> Rumors of intentional infections have usually proven false.

10:17 AM · Jul 22, 2020 · Twitter Web App

Bachman @ElonBachman

In life years lost, a hunger death is worth around ~70 COVID-19 deaths

So in a single month, lockdowns reduce life by more than COVID-19 reduced it in eight months.

We were saying this in Feb/March, and being called "Grandma killers"

> **The Spectator Index** @spectatorindex · Jul 27, 2020
> JUST IN: UN estimates that up to 10,000 children are dying a month due to hunger caused by coronavirus restrictions

9:17 AM · Jul 28, 2020 · Twitter for Android

John Ziegler @Zigmanfreud

Number of Californians under the age of 18 who have died in 2020 due helicopter accidents on their way to a basketball game: 3

Number of Californians under the age of 18 who have died in 2020 due to a coronavirus: 0

#Perspective

7:14 PM · Jul 19, 2020 · Twitter for iPhone

Michael Tracey @mtracey · Jul 21, 2020
Spare me your anguished pleas for people to follow pandemic protocols if you have no issue with these ongoing non-socially distanced mass gatherings -- at which people chant directly into one another's faces -- in cities were **COVID** cases are spiking

TEAM REALITY

Justin Hart @justin_hart

This is bad. And we did it to ourselves.

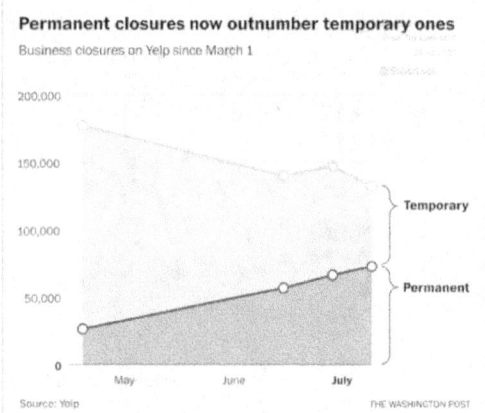

3:05 PM · Jul 28, 2020 · Hootsuite Inc.

Phil Kerpen @kerpen

"I don't think I can emphasize it enough, as the director for the Centers for Disease Control, the leading public health agency in the world: it is in the public health interest that these K-12 students get the schools back open for face-to-face learning."

7:36 PM · Jul 31, 2020 · Twitter Web App

Kyle Lamb @kylamb8

So in South America, they're taking credit for the mitigation measures keeping the flu away but it hasn't stopped them from getting hit with 3.5 million cases or 100,000 deaths. Huh. Interesting logic. Did these geniuses consider Covid "all but" kept the flu away? lol

7:43 PM · Jul 23, 2020 · Twitter Web App

Wes Pegden @WesPegden

The U.S. has committed innumerable and gravely consequential blunders in how it has chosen to respond to the COVID-19 pandemic.

Future generations will look back on the wide resistance to maintaining school for children as the among the most shortsighted and devastating of all.

8:29 PM · Jul 15, 2020 · Twitter Web App

Justin Hart @justin_hart

Dr. Fauci throwing out the first pitch as a symbol of our government's prowess and capability to address #COVID19 and.... doh!

7:08 PM · Jul 23, 2020 · Twitter for iPhone

Phil Kerpen @kerpen

This is evil. Based on the politics of envy and in the absence of any valid health rationale, Montgomery County is ordering private schools that comply with all health guidelines to close. Sacrificing children's wellbeing on the altar of politics.

8:18 PM · Jul 31, 2020 · Twitter Web App

Mark Changizi @MarkChangizi

Large variability in social distancing ...

No variability in death curve.

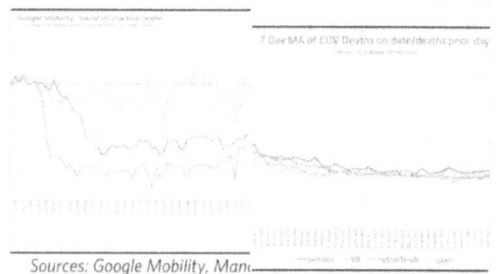

11:38 AM · Jul 13, 2020 · Twitter Web App

Pajamas It Is
@HeckofaLiberal

THREAD: Let's clear some stuff up. It doesn't matter if it's on Twitter or Facebook, but the past several days I have been hearing the SAME EXACT narrative come out of the discussions I've had with dozens of people over multiple platforms. 1/

9:03 PM · Jul 5, 2020 · Twitter for iPhone

None of them have any relation to each other, and some of them aren't even in the same state or sometimes the same country, and the same parroted talking point is coming out. Goes something like this... ++ "oh well the deaths aren't even the most important thing this virus is new and we know people can have longterm effects from this virus and they may never fully recover." STOP SAYING THAT!!! It's not true and none of the data bears out anything even remotely like that. ++ This must be the new narrative because everyone is saying the same thing. The one which gets everyone to just put on that mask! STOP IT! Can this virus cause long term effects? Maybe. It may be possible but it's also possible for other viruses to do the same including the flu. ++ A bout with a cytokine storm is not good and it can leave people suffering for a very long time. This is NOT common though regardless of the illness. That's why you don't know about it. Guess what? It's also not common with SARS-cov2. ++ Definitely some people can be affected that way but the VAST MAJORITY of people have mild to no symptoms and recover fully in a couple of weeks. As deaths are still dropping even as cases "rise" this new narrative gets people to stay afraid. ++ And no, people are not walking around with "glass-like opacity in their lungs." There was a small study which found some people who were not showing symptoms had lung inflammation that the researchers did not think was permanent and more akin to walking pneumonia. ++ And because we don't check asymptomatic people with other respiratory illnesses, the flu is mostly asymptomatic, we don't know if this also happens with other illnesses. So please, stop passing along nonsense. We know more about this virus every day. ++ We will learn more as we go along but passing on a narrative just to push other narratives is simply wrong. All you do is scare people. If an article says "experts are increasingly saying" disregard it. ++ If they can't name the study and link to it, it usually isn't true, and often when they do link to it, the actual study doesn't say what the article is telling you it says. And that's not good news. 10/end

 Prof Francois Balloux ✓ @BallouxFrancois · Jul 11, 2020
Children have to go back to school soon if we hope to avoid major long-term societal damage and further inequality. Young children are at no risk and play a tiny role in #SARSCoV2 transmission. European countries reopened schools in May with no uptick in cases (e.g. Switzerland).

Gummi Bear
@gummibear737

I order food for delivery at least once a day.

Ever since I realized "Covid on everything" is a psychosis I started 2 support local restaurants. You are not getting Covid from ordering in

important we support the small businesses in our communities

Don't fear delivered food

6:05 PM · Jul 11, 2020 · Twitter for iPad

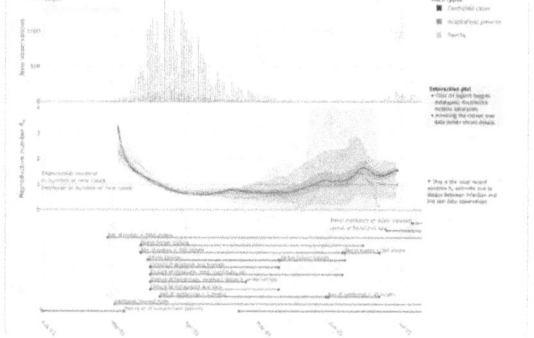

TEAM REALITY

Phil Holloway™
@PhilHollowayEsq

So Georgia kids can't be in school - but for only $800 per month extra, they can be in the extended all-day after school program.

8:45 AM · Jul 29, 2020 · Twitter Web App

Martin Kulldorff
@MartinKulldorff

While #COVID19 deaths are counted and visible in the present, the missed health care visits during the #lockdown will lead to deaths that are spread out over many years to come.

3:35 PM · Jul 29, 2020 · Twitter Web App

Prof Francois Balloux
@BallouxFrancois

I suspected the level of fear/stress was concerning, but hadn't anticipated people were literally dying from it. Cohort study in the US finds 7.8% of stress cardiomyopathy during the #COVID19 pandemic, compared with pre-pandemic incidences of 1.5% -1.8%.
jamanetwork.com/journals/jaman...

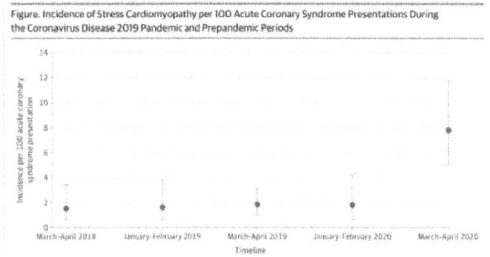

11:19 AM · Jul 11, 2020 · Twitter Web App

Show Me The Data
@txsalth2o

Why does @MartinKulldorff only have 500ish followers?

Professor Harvard Medical School. Disease surveillance methods. Infectious disease outbreaks. Vaccine safety.

Fighting for our kids with data. Please follow this man and share his wisdom.

6:57 PM · Jul 23, 2020 · Twitter Web App

Wes Pegden
@WesPegden

Suggestion:

Before saying: "look at that country, they had X cases in teachers, so we should keep schools closed."

Please look up:
Y: total # teachers
A: total # cases among adults
B: total # adults

and as a basic first pass before posting, consider comparing X/Y with A/B.

3:12 PM · Jul 23, 2020 · Twitter Web App

AJ Kay
@AJKayWriter

Let's be clear:

People wear masks to avoid moral judgement, not to "protect others".

12:35 PM · Jul 16, 2020 · Twitter Web App

Woke Zombie
@AWokeZombie

Day # 128 of "Flattening the Curve" in NJ. For the record, we still cant eat inside, cant go to the gym, still stuck in Phase 2 on indefinite "Pause". Our governor farms old deaths as new, can't do math, and has broken 4 of his own EOs. #ReOpenNJ

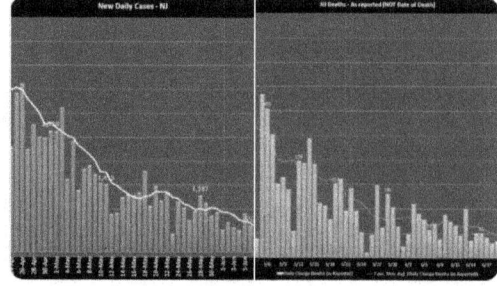

8:39 PM · Jul 20, 2020 · Twitter Web App

TEAM REALITY

Vinay Prasad, MD MPH
@VPrasadMDMPH

Today on social media, I witnessed a physician sharing the obituary of a young person & a tweet of the same person a few weeks ago being critical of masks... shaming the dead

My God, has ethics and decency and compassion vanished?

[thread]

3:49 PM · Jul 11, 2020 · Twitter Web App

No physician would dare shame a person who died of lung cancer by sharing a photo of that person smoking in an effort to curb cigarette use. It would be despicable, loathsome, & worse; no matter the 'cautionary lesson' and yet somehow this idea has vanished for covid19 ++ I understand many are anxious or concerned, and many believe a variety of interventions make sense. But please, doctors do not blame our patients for disease. And, don't reply to this thread saying 'this is different' ++ It is not different. It is so shameful to mock the dead, in a polarized world with unclear messaging, and uncertainty. A new low for this website.

Martin Kulldorff
@MartinKulldorff

Important Scandinavian study on #schoolsreopening by @Folkhalsomynd and @THLresearch: "Closing of schools had no measurable effect on the number of cases of #Covid_19 among children" & "no increased risk for teachers."
folkhalsomyndigheten.se/contentassets/...
@HNohynek

11:58 AM · Jul 12, 2020 · Twitter Web App

AJ Kay
@AJKayWriter

Next time someone says, "What's so hard about wearing a mask??? Jeez..."

Reply with, "Ok. When can we stop?"

The ensuing answer will be an ad hoc justification with no substance...but perhaps you'll make them think about the implications a little more.

12:29 PM · Jul 25, 2020 · Twitter for iPhone

Bethany S. Mandel
@bethanyshondark

Receptionist at my doctor is quitting in August so she can homeschool her kids because schools aren't reopening. Keeping schools closed forces women out of the workforce.

11:35 AM · Jul 28, 2020 · Twitter for iPhone

Prof Francois Balloux
@BallouxFrancois

The CDC released revised estimates for Infection Fatality Rates two days ago. Their most plausible estimate now stands at 0.0065 (0.65%). While the CDC is not immune to political pressure, I do not believe they would 'make up' numbers.
cdc.gov/coronavirus/20...

Gummi Bear
@gummibear737

Almost 50% of deaths are in nursing homes

The average length of stay before death in a nursing home is 14 months

Children on the otherhand have 70-80 years of life ahead of them. They're not being schooled. Their parents are losing their jobs and health insurance.

This is nuts

1:43 PM · Jul 19, 2020 · Twitter for iPad

Parameter	Scenario 1	Scenario 2	Scenario 3	Scenario 4	Scenario 5: Current Best Estimate
R_0*	2.0	2.0	4.0	4.0	2.5
Infection Fatality Ratio, Overall†	0.005	0.005	0.008	0.008	0.0065
Percent of infections that are asymptomatic§	10%	70%	10%	70%	40%
Infectiousness of asymptomatic individuals relative to symptomatic¶	25%	100%	25%	100%	75%
Percentage of transmission occurring prior to symptom onset**	35%	70%	35%	70%	50%

7:39 AM · Jul 12, 2020 · Twitter Web App

TEAM REALITY

Hold2
@Hold2LLC

Someone asked how our Worldwide Hope-Simpson chart would look just with US states, so we put it together. Can't remember who it was, but thank you for the idea.

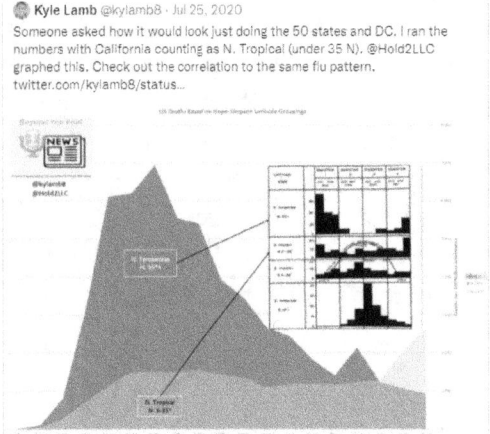

11:46 AM · Jul 26, 2020 · Twitter Web App

Kyle Lamb
@kylamb8

Seasonal pattern of Covid-19 deaths compared to embedded chart of 1964-75 flu outbreak by four seasons. Countries & states split into 35+ N. Lat, 0-35 N, 0-30 S and 30+ S. Flu is the same except 0-30 and 30+ North. This is a striking similarity. Match this with left half of Flu.

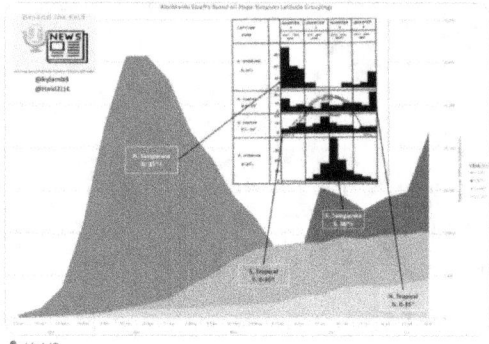

👤 Hold2

2:17 AM · Jul 24, 2020 · Twitter Web App

Emma Woodhouse
@EWoodhouse7

Friendly reminder that 5 of the "COVID-19 deaths" in Cook County (IL) were drug overdoses.

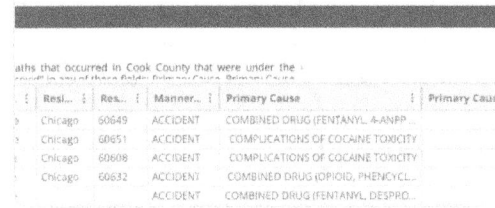

11:41 PM · Jul 21, 2020 · Twitter Web App

Bethany S. Mandel
@bethanyshondark

Has any other essential industry put up a fight about working like teachers have? I never saw groups of doctors, nurses, meat packers, truckers, grocery store workers stand up and say they won't work until there's a vaccine and to hell with our society.

7:14 PM · Jul 14, 2020 · Twitter for iPhone

Prof
@covidtweets

We cannot hostage kids and ourselves indefinitely. I understand the early concern about the healthcare system being overwhelmed. That was based on grossly inaccurate models and is apparently not happening. Why are we still not shifting the responsibility to individuals?

11:17 PM · Jul 31, 2020 · Twitter Web App

Gummi Bear
@gummibear737

Life expectancy (LE) in 1900 - 47. Penicillin discovered in 1928 - LE was 57. Now LE is 80.

Yet we are so scared of a virus that kills 0.4% (I stand by this number), that we destroy our society out of fear?

"we are not descended from fearful men" Edward R. Murrow

Just sayin

5:05 PM · Jul 18, 2020 · Twitter for iPad

John Ziegler
@Zigmanfreud

I'm not certain about the science on these, but I'm VERY confident if at start of this fiasco Trump had demanded:

-Masks mandated
-Schools 100% online
-Hydroxychloroquine banned
-Censoring opposing views

That the woke crowd would be 100% against all 4!!
newsweek.com/key-defeating-...

11:45 PM · Jul 27, 2020 · Twitter for iPhone

TEAM REALITY

Emma Woodhouse
@EWoodhouse7

It's a sad day when the teachers union in the largest K-8 district in Illinois is promoting fear by dressing in black, donning masks, and caravan thru their students' neighborhoods as harbingers of doom, gloom, & death, to promote a not-so-brave new world of schooling. 😞

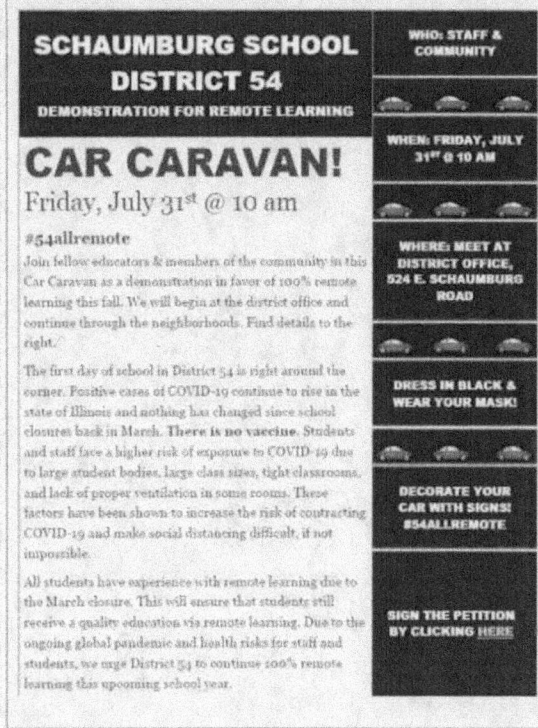

2:20 PM · Jul 27, 2020 · Twitter for iPhone

Bethany S. Mandel ✓
@bethanyshondark

Here's what teacher's unions want: the pay, benefits and recognition of being an essential part of the workforce, but none of the responsibilities. Pick one. You're either essential or you're not. If you believe you're not, if you think education is optional, choose a new career.

9:08 AM · Jul 22, 2020 · Twitter for iPhone

Kyle Lamb
@kylamb8

Healthy young adults testing positive for a virus, which does not indicate an active infection, that is giving them zero symptoms, puts them in little or no threat of harm, and after four months people freak out like they've been diagnosed with Ebola. Hysteria.

11:18 AM · Jul 27, 2020 · Twitter Web App

Eric
@The_OtherET

Sweden #COVID deaths per 100K is 43.9, now near zero: failed experiment.

New York COVID deaths per 100K is 168, now at "zero" (it isn't, data lag): magnificent success story of competent government. @JRubinBlogger and her ilk are all hacks.

11:00 PM · Jul 12, 2020 · Twitter for iPhone

Erich Hartmann
@erichhartmann

Replying to @EthicalSkeptic

LockDown is a strategy based on control and force

2:59 PM · Jul 25, 2020 · Tweetbot for iOS

Kyle Lamb
@kylamb8

Remember when early in March it was said you were supposed to listen to the experts? Today, both @twitter and @facebook censored actual licensed physicians speaking on Covid-19 under the pretense it was "false information."

Listening the experts was always a lie.

12:40 AM · Jul 28, 2020 · Twitter Web App

Zac Bissonnette
@ZacBissonnette

This seems like a bad way to make sure young people don't give covid to older people.

Did anyone think about the flow through of these policy decisions for, say, 4 seconds?

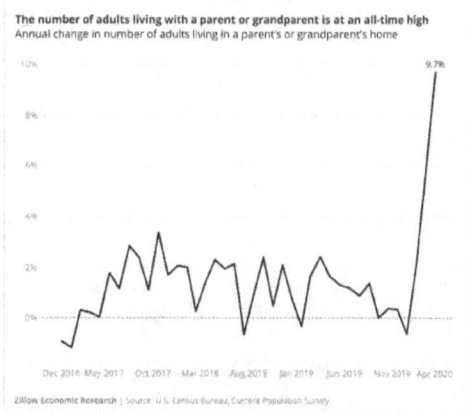

12:02 PM · Jul 5, 2020 · Twitter for iPhone

AUGUST 2020

AJ Kay
@AJKayWriter

There's a non-zero chance that if the financial incentives to code #Covid_19 were halted tomorrow, the U.S. "crisis" would end very shortly thereafter.

11:59 AM · Aug 4, 2020 · Twitter for iPhone

TEAM REALITY

Federico Andres Lois
@federicolois

I am pretty sure that time will show that COVID policies implemented worldwide have scratched (if not passed) the threshold to become the largest unethical medical experiment of all time. What's unethical to me? Putting the vulnerable at more risk than non-vulnerable.

9:45 PM · Aug 8, 2020 · Twitter Web App

Bethany S. Mandel ✓
@bethanyshondark

The goal was to flatten the curve. It was never to stop anyone from getting this virus ever. That is not possible. The genie is out of the bottle.

9:40 AM · Aug 2, 2020 · Twitter for iPhone

Gummi Bear
@gummibear737

I am becoming increasingly convinced that the actual number of people who have been infected is far larger than what we are able to measure via seroprevalence

The question is by how much

7:25 AM · Aug 23, 2020 · Twitter for iPhone

Victoria Fox
@drvictoriafox

No one's suggesting the discovery of robust T-cell responses means we should all go out licking each other. But It is good to see we've moved on from the odd notion there can be no immunity to COVID, to a more sensible debate that correlates with known immune mechanisms.

3:12 AM · Aug 20, 2020 · Twitter for iPhone

Stefan Baral ✓
@sdbaral

A realization which may or may not be worth sharing.

#COVID19 may represent the largest risk transfer by income in recorded history.

We bought ourselves out of acquisition risks by transferring those risks onto those who kept society afloat.

And they have paid with their lives

12:01 PM · Aug 28, 2020 · Twitter Web App

Real Developments
@pdubdev

The saddest lunchroom ever.

Only selfishly scared people would inflict this insanity on children.

This is not about kids but about the hysterical fear of adults.

Child abuse.

11:03 AM · Aug 18, 2020 · Twitter for iPhone

Victoria Fox
@drvictoriafox

The COVID-19 pandemic has really changed me. I've seen the media in a whole new light. I can't read the news anymore without wondering is this another topic they got wrong - like COVID. NPR now infuriates me. I'm not sure I will be able to believe news reports for a long time.

1:38 AM · Aug 22, 2020 · Twitter for iPhone

John Ziegler ✓
@Zigmanfreud

There are lots of indications that masks are actually harmful. Heck, look at the data in LA and CA after the mask mandates.

But more importantly, I am against the government mandating them. If they can do that to us, what can't they do? And when will the mandates ever end now?!

10:45 PM · Aug 12, 2020 · Twitter for iPhone

TEAM REALITY

Justin Hart
@justin_hart

More hellish delusions.

8:49 PM · Aug 19, 2020 · Twitter Web App

Phil Kerpen @kerpen

🚨 The CDC no longer recommends asymptomatic testing, even post-exposure. 🚨
cdc.gov/coronavirus/20...

12:30 PM · Aug 24, 2020 · Twitter Web App

Unfortunately, this lasted for about a week. Facing immense pressure from the usual suspects, the CDC caved and walked this back. Prof

Kyle Lamb @kylamb8

If you love hypocrisy, remember the same people crying that the CDC changed its testing recommendations calling it political are the ones that bullied the CDC into changing its 0.26% IFR. They also shamed WHO into walking back "rare" asymptomatic transmission. Got to love irony

1:53 PM · Aug 26, 2020 · Twitter Web App

Vinay Prasad, MD MPH @VPrasadMDMPH

I absolutely do not want the FDA to approve an unproven vaccine under political pressure

That said: anyone who thinks the FDA has historically been resistant to political pressure and only "follows the science", does not know anything about the FDA.

4:42 PM · Aug 22, 2020 · Twitter Web App

Justin Hart
@justin_hart

We've lost our minds.

 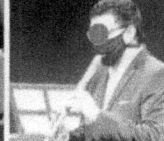 

5:14 PM · Aug 25, 2020 · Twitter for iPhone

Phil Kerpen @kerpen

CDC: 25.5% of Americans age 18 to 24 seriously considered suicide in the past 30 days. 16% of age 25 to 44.
cdc.gov/mmwr/volumes/6...

3:17 PM · Aug 13, 2020 · Twitter Web App

Pajamas It Is @HeckofaLiberal

I want to know why @JoeBiden is all about mask mandates but has not said one word about Vitamin D intake and Americans eating healthier to stop the spread? Not one PSA about health.
thelancet.com/journals/landi...

3:10 PM · Aug 16, 2020 · Twitter for iPhone

Vanessa @vtal42

Honest question- If we are going to restrict teens from going to school, why aren't they restricted from working? I see them all over- restaurants, gyms, babysitting and tutoring to fill the void of schools. Are we really worried about them, or is it just the teachers?

6:25 PM · Aug 21, 2020 · Twitter Web App

TEAM REALITY

Eli Klein
@TheEliKlein

Today @CDCgov confirmed Covid immunity lasts AT LEAST 3 months after recovery.

Our news promptly misrepresent this, claiming immunity ONLY lasts for 3 months.

Maddening.

6:26 PM · Aug 14, 2020 · Twitter for iPhone

Hold2 @Hold2LLC · Aug 20, 2020
Whoa...I need to transcribe what this, but here's the summary.

She said 2 critical things:

1) Anyone with a COVID+ test within a certain timeframe of death is counted as a COVID death

2) Anyone with COVID _anywhere_ on the death certificate is counted as a COVID death

Emma Woodhouse
@EWoodhouse7

If your state is defining a COVID death as the death of any person who tested positive for COVID w/in 30 days (IL) or 60 days (AZ) - & the deaths = $$$ - then of course your state wants to test anyone & everyone. It allows them to capitalize on more deaths, incl predictable ones

4:24 PM · Aug 24, 2020 · Twitter for iPhone

Martin Kulldorff
@MartinKulldorff

It is absolutely stunning to observe how the scientific community has reacted to the public health aspects of the pandemic. When the fog clears, one of the consequences of the pandemic will be public distrust in science and scientists.

9:41 AM · Aug 12, 2020 · Twitter Web App

Prof Francois Balloux
@BallouxFrancois

>1M mostly healthy people are killed on the world's roads each year. There's no call for 'zero risk', and road safety measures aim to minimise harm while allowing for the seamless flow of people/goods. Pre-#COVID19, epidemiology felt similarly pragmatic and non-ideological to me.

4:44 PM · Aug 21, 2020 · Twitter Web App

Wes Pegden
@WesPegden

As schools open around the world, U.S. media are naturally discussing the experience of many of the European countries shown below.

2:48 PM · Aug 26, 2020 · Twitter Web App

AJ Kay
@AJKayWriter

The data is starting to come in about the cost of lockdowns in terms of human life.

TLDR: CDC recorded 10K excess deaths from other causes vs 8K of those w/ COVID in 25-44 year olds since March—presumably from suicide, despair, overdoses, and violence.

2:57 PM · Aug 21, 2020 · Twitter for iPhone

ClownBasket
@ClownBasket

Replying to @covid_clarity

Ha. There is no difference between the politics and "science" right now. Earlier this week the CDC changed its testing protocol. But then got blowback so they walked back their change a day later. The change didn't fit the political narrative. Fear and control. That's the game.

2:37 PM · Aug 30, 2020 · Twitter for iPhone

Woke Zombie
@AWokeZombie

So how's schools going in Georgia? Been eerily quiet since the mass quarantine for the few positives. How did the other 1,700,000 students make out? Its been 2+ weeks since that story.

10:42 PM · Aug 28, 2020 · Twitter Web App

TEAM REALITY

Prof Francois Balloux
@BallouxFrancois

Actually, I reckon 'Science' will come out stronger on the other end of the pandemic. Many excellent scientists were completely unprepared to deal with the fear and anger the pandemic unleashed, and found it difficult to get their voices heard. I feel this is slowly changing ...

> **Gummi Bear** @gummibear737 · Aug 26, 2020
> Replying to @BallouxFrancois
> How does science recover from what's happened during Covid?
>
> Scientific debate was shut down
>
> All the wrong science was used to drive policy
>
> "Celebrity Scientists" sacrificed their integrity to pander for likes and followers
>
> Is there even going to be a recognition of this?

4:40 PM · Aug 26, 2020 · Twitter Web App

Prof
@covidtweets

Inspired by @Hold2LLC

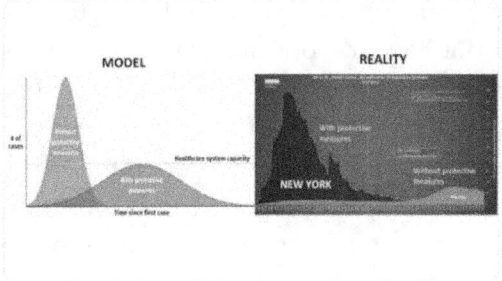

12:56 PM · Aug 26, 2020 · Twitter Web App

Mark Changizi
@MarkChangizi

"But do we have the right to suspend civil liberties in this way?" — A quote from no government official anywhere in 2020

10:16 AM · Aug 5, 2020 · Twitter Web App

Hold2
@Hold2LLC

"Forcing the tradeoffs to happen out of sight."

This is exactly the issue I've been harping about. There is an invisible equation where only one side is being measures...the COVID deaths side. The other side is being routinely ignored because those problems happen "out of sight"

> **Bachman** @ElonBachman · Aug 23, 2020
> Replying to @ZacBissonnette
> It's a real world version of the Trolley Problem
>
> People don't want to be responsible for confronting tradeoffs regarding human life, so they support ultra Puritan policies that force the tradeoffs to happen out of sight
>
> They close their eyes and imagine reality disappears

4:20 PM · Aug 23, 2020 · Twitter Web App

Kyle Lamb
@kylamb8

Fauci said before mid March:

He wasn't concerned about the virus

No reason to shut down

Masks didn't help, weren't needed

Asymptomatic spread was rare and didn't drive a virus

Either the guy is not good at his job or he was right before he got political

11:22 PM · Aug 24, 2020 · Twitter Web App

Ann Bauer
@annbauerwriter

#Obesity is a prime risk factor for death from #COVID19. Shutting pp in their homes, giving them nothing to do but eat, limiting their exercise and sun, contributes to obesity. If I can understand/anticipate this, why can't our epidemiology gurus?

10:14 AM · Aug 21, 2020 · Twitter Web App

Emma Woodhouse
@EWoodhouse7

This morning, a group of 15 Cook Co residents - me included myself - met in Fountain Sqaure in downtown Evanston to share our concerns about what's happening in our state with Covid19

We did not wear masks, & we were not sitting 6 feet apart

Onlookers stared

It's a start. 💥

12:04 PM · Aug 22, 2020 · Twitter for iPhone

AJ Kay
@AJKayWriter

I'm out of state this weekend for a wedding.

People gathering, sharing,
laughing, hugging, sharing food & drinks. Dancing. No artificial distance. No performative safety. No fear of friends & loved ones.

Beautiful

The "new normal" will never be normal. Humans need connection.

1:50 PM · Aug 29, 2020 · Twitter for iPhone

Eric
@The_OtherET

Good morning @RexArcherMD, couldn't help but notice you still have this tweet up saying Kansas City should be more like NYC. If Kansas City were like NYC, we'd have 1,385 #COVID19 deaths. We currently have 96 in KCMO. But at least their curve is flat now!

12:25 PM · Aug 28, 2020 · Twitter Web App

But there isn't just COVID mortality - what about from lockdowns? If Queens is any indication, suicides in NYC have spiked drastically since 2019 - they saw the same number of suicides in 6 weeks that they had previously seen over 4 months in 2019. ++ Not just suicides, but drug overdoses too - Queens has had a 56% increase in overdose deaths compared to 2019. ++ 1/4 of apartment renters in NYC have not been able to pay their rent since March. That's 1.4 million people that could potentially lose their homes. ++ 1/3 of NYC small businesses are likely to close forever. I for one like our small businesses in KC - one in particular, Tapcade, has been one of my favorite hangouts for years. They permanently close on Sunday. ++ In June (cases and deaths were essentially zero at this point in NY) lockdowns were costing NYC $173 million per day. But I'm sure they were just suppressing the "next wave"! ++ And to make things worse, people are leaving NYC in droves, bringing their tax dollars with them. How are they going to make up that kind of deficit? ++ But hey, let's look at our situation in KC - as you obsess over rising case numbers, deaths have flattened. Case numbers have gone up, and deaths have gone down. Deaths started slowing in June/July, which is when KC mobility had nearly reached pre-pandemic levels. Huh? ++ So @RexArcherMD, which part of NYC do you want us to be like? The huge number of COVID 19 deaths? The huge rise in suicides/overdoses? Bringing a tsunami of impending homelessness to 1/4 of KC renters? Destroying small businesses? Bankrupting the city?

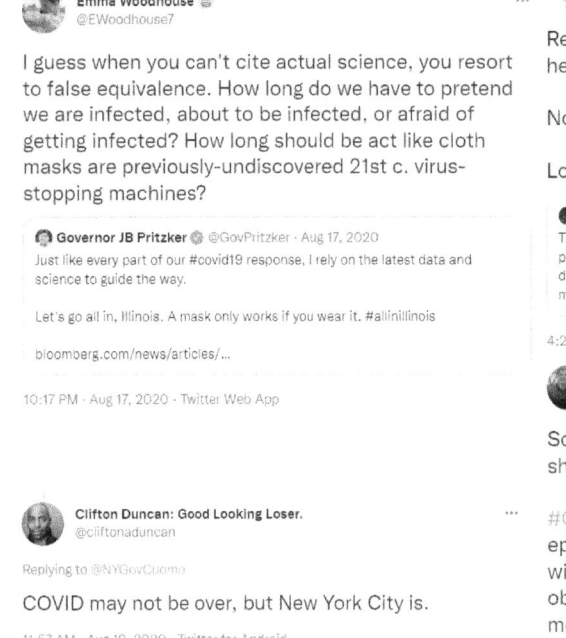

Eric
@The_OtherET

So the German JAMA #COVID19 cardiology study that has been referenced thousands of times to shut down the #Big10 and #PAC12 seasons by media including @ESPN presents data that is statistically impossible. Basically everything in this table below is bogus. Let me explain:

11:31 AM · Aug 18, 2020 · Twitter Web App

For this example I'm going to use systolic blood pressure but it can be anything that was highlighted above by @ProfDFrancis. In the table above, the parentheses mean they found an interquartile range for systolic BP of 125-133 - or 50/100 of their subjects fell in this range. ++ This is a problem. In a much larger study across the German population the IQR is much larger. For example, if we take the 41-50 age range from this larger study (similar range to the JAMA study) they found a mean of 130.5 and a standard deviation of 16.8. ++ Assuming a normal distribution curve (which I think is a safe assumption) you have about a 19% chance of finding someone in Germany with a systolic blood pressure within this range. ++ Again, going back to the JAMA study they said they found roughly 50 people out of 100 that were within the 125 - 133 mmHg range for systolic blood pressure. But based on the larger German study, you would expect to only find 19 people out of 100. So what are the chances of 50? ++ Well…assuming a Poisson Distribution (again I think a fair assumption) this is what the curve looks like - this means while finding around 19 people is the most likely, there are other feasible outcomes as well. But where is 50? It doesn't even warrant a place on the x axis. ++ Again, the JAMA COVID cardiology study is full of statistics like this - they are 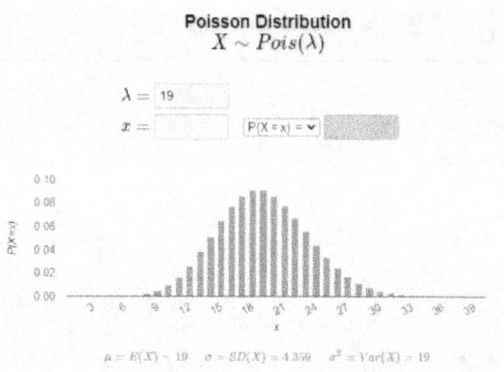 either wrong, a result of selective bias, or worse, completely made up. I will gladly admit fault if any of my numbers are wrong here - but they follow pretty basic statistical theory. ++ Shamelessly tagging @ClayTravis, @kylamb8, @justin_hart, @AlexBerenson, @kerpen to hopefully read/RT this thread to bring it to light - and have smarter people than me check my work. But if it's right - they need to retract this study immediately.

Martin Kulldorff
@MartinKulldorff

"We should appreciate young adults who help generate herd immunity by living normal lives and keeping society afloat. When people throw misguided complaints at you, falsely claiming that you are endangering others, remember that the opposite is true."

12:04 PM · Aug 17, 2020 · Twitter Web App

Zac Bissonnette
@ZacBissonnette

As a longtime Democrat, I gotta say:

I can forgive a lot, but the decision by most Democrats to promote baseless terror about schools as covid hot zones is probably the most disgusting thing I've ever seen done by people I'd thought were decent.

6:24 PM · Aug 23, 2020 · Twitter for iPhone

TEAM REALITY

SEPTEMBER 2020

Jennifer Sey ✓
@JenniferSey

Most private schools in SF will open on Monday. Public schools won't. My kids will remain at home. I'm sad for them and their peers. I'd like to understand the plan to address how far behind many will fall. Is there one? Or do we pretend screen learning works just fine?

3:16 PM · Sep 18, 2020 · Twitter Web App

TEAM REALITY

Mark Changizi
@MarkChangizi

The Covid pandemic was a litmus test.

This was THE moment in several generations to recognize the treacherous signs of groupthink and authoritarianism.

How did you fare?

11:31 AM · Sep 27, 2020 · Twitter Web App

When history came knocking, telling you that in order to save society you have overturn it, suspend liberties, and fight to suppress the deniers, did you do as instructed? ++ When everyone you knew, and every news outlet you watched, told you that the sky is falling and that everyone must fall in line, abandon normality, civil rights, and livelihoods, THAT was the most important time in your life to stand up and be independent. Were you? ++ When a new danger approached, requiring society to work together as one in order to defeat it, did you help pull society together? Or did you, instead, ask whether acting to "pull society together" might in actuality amount to authoritarian bullying, informing and snitching? ++ The narrative is clear. The science is settled. Not one news story indicates an ounce of skepticism. When the implications of such consensus have unprecedented societal implications, did you dig deeper, examine the evidence yourself, and look for any alternative viewpoints? ++ This was a test. But it was more than a test. ++ ... ++ Passing this litmus test, and being the voice against the mob, has its risks, of course, which is why folks keep their mouths shut.

Real Developments
@pdubdev

New study shows .26% Infection Fatality Rate exactly as Stanford genius John Ioannidis calculated (and was vilified for) back in March.

That's one smart dude

acpjournals.org/doi/10.7326/M2...

1:36 AM · Sep 3, 2020 · Hootsuite Inc.

Victoria Fox
@drvictoriafox

I've come to realize every scientist has missed a critical point in the suppression/immunity debate. Pandemics don't end with suppression or immunity. They end when people say they do. When people decide they're done being scared & accept the risk of resuming their lives.

12:52 PM · Sep 30, 2020 · Twitter for iPhone

Eli Klein
@TheEliKlein

I'm a lifelong Democrat

Watching red states handle Covid infinitely better than blue states is a nightmare

The NY (blue / closed economy, higher death rate) to FL (red / open economy, lower death rate) comparison is generally applicable across the US

Eli Klein @TheEliKlein · Sep 14, 2020
Florida's herd immunity strategy is in many ways more robust than Sweden's.

Watching Florida crush Covid with population immunity is phenomenally powerful and interesting.

It's a shame our "once credible" media and experts are ignoring it.
Show this thread

8:15 AM · Sep 18, 2020 · Twitter for iPhone

Eric
@The_OtherET

So literally the day after emails were exposed showing they covered up data showing bars and restaurants didn't contribute to #COVID19 spread, Nashville:
- Allowing fans at Titans games
- Bars open at 50% capacity
- High school football can hold games

Unbelievable.

11:34 AM · Sep 17, 2020 · Twitter Web App

TEAM REALITY

Don Wolt @tlowdon

Fauci is like a French general in 1940 urging the building of a deeper, stronger Maginot Line even as the Germans sweep around it. He's fixed on a single theory of defense & appears incapable of updating his strategy as new facts emerge. We need a more agile, dynamic thinker.

1:11 PM · Sep 23, 2020 · Twitter Web App

John Ziegler @Zigmanfreud

It has now been over 2 months since Sweden recorded even double digit (1 for every million people) deaths with/of COVID. This is with life there near normal (0 masks!).

The total lack of media/expert curiosity regarding why/how the hell this can be is both staggering & expected!

6:36 PM · Sep 18, 2020 · Twitter for iPhone

Victoria Fox @drvictoriafox

Seriously, what kind of dick to you have to be to cancel Halloween. As activities go, trick-or-treating doesn't get any safer. It's Outdoors. Involves kids (low risk spreaders) in Halloween consumes (i.e. masked). If you're still scared turn off your light & don't participate.

10:53 PM · Sep 9, 2020 · Twitter for iPhone

Angry Cardiologist @AngryCardio

SARSCoV2 as a cause of widespread myocarditis is pure, unadulterated FUD.

Everything is "possible" or "a sign of". Nothing substantial. Nothing that hasn't been discussed ad nauseum. Thanks, @nytimes!

7:52 AM · Sep 24, 2020 · Twitter for iPhone

Prof Francois Balloux @BallouxFrancois

'Zero COVID19' would have implied global elimination of the virus, anywhere in the world. This window of opportunity was gone by early February and possibly earlier, and definitely not over the summer. I do not believe it is helpful to cry over spilt milk, anyway.

11:57 AM · Sep 13, 2020 · Twitter Web App

Vinay Prasad, MD MPH @VPrasadMDMPH

It is painful it is to watch people who I used to think were careful thinkers take absolute positions on in-person school re-openings that completely ignore the catastrophic downsides to prolonged closure on a generation of kids, particularly poor kids, particularly young kids

7:02 PM · Sep 16, 2020 · Twitter Web App

John Ziegler @Zigmanfreud

The grand USA Experiment officially ended the moment a government-forced shutdown, supposedly to "flatten the curve," became endless NOT because COVID was worse than feared, but because it was NOT as bad as predicted & those who messed up became invested in not having been wrong.

6:13 PM · Sep 22, 2020 · Twitter for iPhone

Stefan Baral @sdbaral

I am not sure who needs to hear this, but:

Human beings are more than just vectors of disease
Populations have needs other than #COVID19 prevention
Equity is a foundational principle of public health

Oh yeah, human beings are more than just vectors of disease.

2:26 PM · Sep 8, 2020 · Twitter Web App

Real Developments @pdubdev

Cancer deaths are now skyrocketing above normal levels.

These deaths are a direct result of the lockdowns.

This is a human rights crime. People who could have been treated were denied care, because our government's hysteria.

Source: CDC Weekly Deaths by Cause

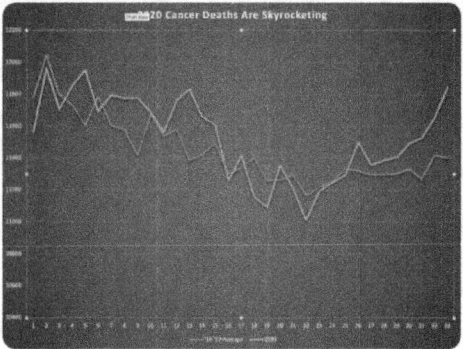

4:16 PM · Sep 4, 2020 · Twitter Web App

Prof @covidtweets

Week 8 of in-person teaching. Still didn't die.

Can someone check in on me in two weeks, just in case?

The risk may not be zero, but it is pretty close and certainly not high enough to justify denying a decent education from our future generations.

12:10 PM · Sep 29, 2020 · Twitter for Android

TEAM REALITY

Martin Kulldorff
@MartinKulldorff

Contact tracing, testing and isolation is important against many infectious disease outbreaks, such as Ebola and post-vaccine measles. It is ineffective, naïve and counter-productive against COVID19, influenza, pre-vaccine measles, etc, and by definition, against any pandemic.

1:55 PM · Sep 19, 2020 · Twitter Web App

Stefan Baral ✓
@sdbaral

Replying to @drvictoriafox @rfsquared and 3 others

I would have never imagined the day when saying the word immunity out loud would make someone presume your political leanings.

Marriage Equality--sure. Equal Pay--sure. Universal health care--sure. Human rights--sure. Housing rights--sure.

Immunity---WTF?

1:00 PM · Sep 3, 2020 · Twitter Web App

Pajamas It Is
@HeckofaLiberal

Redfield better hope vaccines work better than masks.

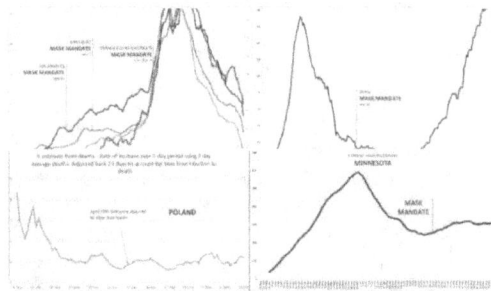

8:23 PM · Sep 16, 2020 · Twitter for iPhone

Kyle Lamb
@kylamb8

This is the IHME model for Sweden, who has not been wearing masks and has fallen to almost no deaths, for projected deaths if...they don't wear masks.

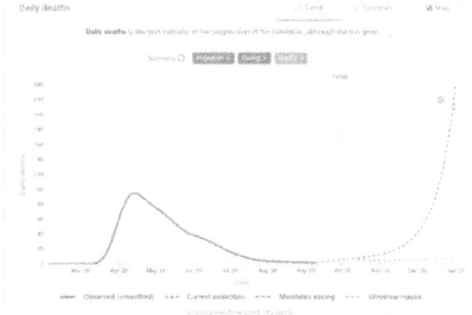

3:46 PM · Sep 29, 2020 · Twitter for Android

Justin Hart
@justin_hart

We've lost our ever-lovin' minds.
h/t @JTUGS

3:20 PM · Sep 27, 2020 · Twitter Web App

Nick Foy
@TheNickFoy

I made a map to simplify the risk/benefit analysis for our school board who maintains that keeping schools closed is smart.

I crunched data locally, regionally, nationally, and internationally to asses the risks from school closures vs. risks from covid. Here's what I found:

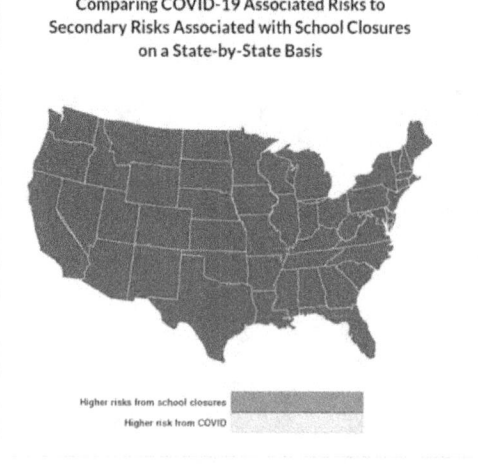

6:23 PM · Sep 27, 2020 · Twitter Web App

TEAM REALITY

Nick Foy
@TheNickFoy

As reported in this morning's @CltLedger, we filed a lawsuit against the CMS Board (et al.) regarding our district's decision to allow for only "virtual learning" this fall.

Here's why we did it: (1/11)

3:45 PM · Sep 8, 2020 · Twitter Web App

As anyone who follows me is well aware, my Twitter feed has become a cesspool of school-related data, commentary, information, etc. I've lost sleep many nights since schools were closed in March. ++ Not because I'm terribly worried about our own kids (we've got resources to find alternative ways to normalize their educational experience), but because of the dramatic effect our public policy is having (and will continue to have) on those without such resources. ++ To reiterate the global data: 1. COVID-19 is far less dangerous to kids than influenza. 2. Kids don't seem to transmit it as well as they do other diseases. 3. Teachers (in places where schools remained open) were at less than average risk of contracting COVID-19. ++ This summer, our nation started to face the reality of systematic oppression and the long-range consequences associated with a lack of social connection, mobility, and resources for people who don't look like me. An indefinite school closure exacerbates these inequalities. ++ Our city recently ranked 50 out of 50 large cities in upward mobility for children born into our lowest income quintile. Missing months of education will significantly worsen these dramatic differences. ++ I'm fortunate to have access to non-financial resources (the internet! friends who are amazing attorneys!) that allow me to be a voice for the 16,000(!) kids in our community who can't even logon for their "remote learning." ++ I'm sure our motives will be (already have been?) politicized, although my reason for párticipating is anything but political. ++ It's entirely data based, honestly assessing the risks of putting kids in schools (seemingly quite low based on millions of data points) with the risks of keeping schools closed (incredibly high). ++ One last thing... As of 9/8/2020, the state of NC has (literally) hundreds of thousands of kids in school buildings. Part-time for public and full-time for private. There are currently 47 cases in school settings. ++ This data matches what we've seen all over the world; that schools are not the primary drivers of transmission accelerating the pandemic.

Martin Kulldorff
@MartinKulldorff

Naïve/dangerous to think all COVID deaths can be prevented. Such strategies increase deaths. We minimize deaths with age-targeted strategy advocated by @SWAtlasHoover and infectious disease epidemiologists like @SunetraGupta @sdbaral @RebeccaChandle1 & me

5:27 PM · Sep 18, 2020 · Twitter Web App

Kyle Lamb
@kylamb8

A health secretary forged a graph to make masks look like they worked. A mayor withheld exculpatory information about bars/restaurants. We've seen suicides, motorcycle accidents, etc. counted as Covid. Individually these are anecdotes. Collectively, they are a pattern of deceit.

12:11 AM · Sep 17, 2020 · Twitter Web App

Gummi Bear
@gummibear737

These Activist Scientists are the reason why people aren't trusting science anymore

They're not even trying to hide their activism anymore

I guess the "science is settled" folks!

Vote for Joe Biden or PEOPLE WILL DIE!

> **Scientific American** @sciam · Sep 15, 2020
> Scientific American has never endorsed a presidential candidate in our 175-year history—until now.
>
> The 2020 election is literally a matter of life and death. We urge you to vote for health, science and Joe Biden for President.
> scientificamerican.com/article/scient...
> Show this thread

9:36 AM · Sep 15, 2020 · Twitter for iPad

TEAM REALITY

Justin Hart
@justin_hart

This is the most wrenching video I've tweeted out during the pandemic - and that includes any of the riot videos. I want you to watch this and know - intimately - the pain these lockdowns have caused. There's a better way.
Please send far and wide.

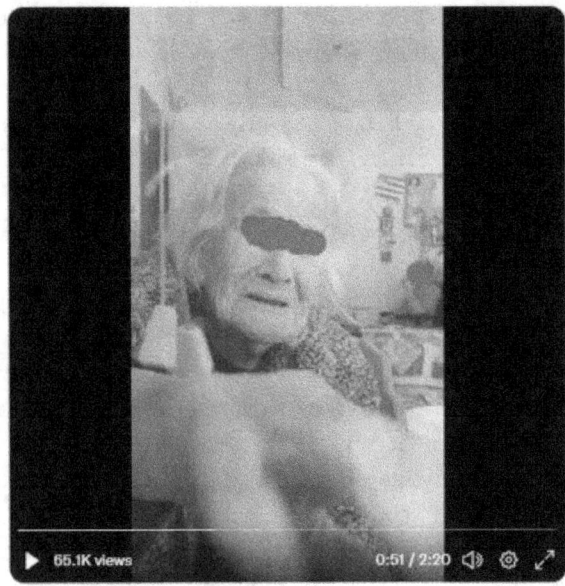

6:18 PM · Sep 28, 2020 · Twitter Web App

The video (https://twitter.com/justin_hart/status/1310705415030628352) shows a nursing home resident, visited by [likely] her son, trying to touch him through the glass, asking him to come inside, then crying in despair...

The way we treated our elderly was just as horrific as we treated our kids during this pandemic.

There WAS a better way...

_prof

AJ Kay
@AJKayWriter

College Math:

Kids at minimal Covid risk normally interacting w/ other kids who are at minimal risk = minimal risk

Healthy kids put into solitary confinement + increased anxiety/guilt/fear + interrupted educations social isolation = significant risk

11:51 AM · Sep 1, 2020 · Twitter for iPhone

Phil Kerpen @kerpen

CDC estimated survival rates by age:
0 to 19: 99.997%
20 to 49: 99.98%
50 to 69: 99.5%
70+: 94.6%

> **Kyle Lamb** @kylamb8 · Sep 22, 2020
> Have you seen the new best estimate infection fatality rates by the CDC? 20-49 is 0.02%. That's a 99.98% survival rate.
> Show this thread

4:09 PM · Sep 22, 2020 · Twitter for Android

Martin Kulldorff
@MartinKulldorff

Tips for scientists engaging in public COVID19 debate
1 Stick to your area of expertise
2 Acknowledge what is known vs not
3 Read opposing views
4 Engage with left and right
5 To gain trust, don't mix in your politics
6 Call out bullying scientists
7 Always polite, always

10:56 AM · Sep 22, 2020 · Twitter Web App

TEAM REALITY

AJ Kay
@AJKayWriter

I have been MIA on Twitter (& publishing), in large part b/c I am a single mom who is trying to both work and homeschool 3 kids, one w/ autism.

Our local district is offering online-only ed at least through Oct & that doesn't work for our fam. So I'm figuring it out..

(1/14)

9:44 PM · Sep 16, 2020 · Twitter Web App

...We are fortunate. We have a safe home, ample food, reliable transportation, and supportive friends...and we are still struggling. All 4 of my girls (one is in college) are stressed, anxious, moody, & tearful. They want to know when this will end & I can't tell them.. ++ It's difficult to combat the message that they'll kill someone if they get too close & that their mere presence in public threatens ppl. Things they've looked forward to for so long: friends, sports, parties, proms, graduation have evaporated & been replaced by a life... ++ ...where the only morally acceptable activity is 'not infecting other people' My HS student (15) has lost almost all of her friends. And not due to distancing, but because we live in a well-off island inside a socioeconomically stressed area & her friends... ++ ...were already impacted by their varying degrees of impoverishment. Now they are home all day unsupervised and while their parents work whatever jobs they can find, they attend "class" with their cameras off, getting high & self-medicating. Of course, they do... ++ ...These kids were doing okay last year. Now I'd be surprised if half of them don't drop out. They're just trying to cope. My college kid took a leave of absence this fall. Not b/c she was scared of Covid, but b/c of the restrictions imposed in the name of "safety"... ++ She was scared of losing her 75K/yr scholarship if she accidentally hugged a friend in public. She wasn't okay being subjected to invasive mandatory testing and arbitrary rules and quarantine when she wasn't sick. She didn't want to live in a pseudo-community... ++ ...which treated its members like vectors, not people, and told them it was their moral obligation to narc on their friends. And it was the right choice. Going away to college isn't about being locked in a dorm room alone, eating alone, and doing homework online alone... ++ It's about being independent for the first time, making mistakes, building relationships, & learning to navigate the world. None of that is happening. My autistic daughter (12) is missing crucial, in-person experiences that are the only reason she's so functional... ++ She is missing her therapies. She is missing the aides who redirected her & kept her on task. She is missing her social skills group. And she has regressed. As for my little one (10), she's just lost and lonely and confused... ++ It feels like her spirit dims a little every day. There are no sports or social programs available that don't require distancing and masks. And I am just ONE mom--with pretty decent circumstances--doing the best I can & it's not enough. I am struggling, too. ++ So my point is this: Shut downs have consequences & the ones carrying the lion's share of the burden are the ones least equipped to do so. And we're being told that it's an immoral position to push back against unnecessary restrictions... ++ ...and play along with calling them "safe"--as if COVID is the only threat to everyone's life or well-being. As if sacrificing the development, education, and well being of a generation is a justifiable cost. I refuse. None of this is "safe" or warranted. ++ People--most importantly, children--are suffering harm and I think pushing back IS the moral position.

Bethany S. Mandel ✓
@bethanyshondark

It's cute how the teachers are complaining about how hard teaching online is, as if they had no alternative. Crazy idea: you should've fought your union to let you do your job the right way. And of all the people I feel badly for in this situation, it's the kids.

9:24 PM · Sep 17, 2020 · Twitter for iPhone

Jennifer Sey ✓
@JenniferSey

Right now in SF I can legally go buy weed in a store (inside) or illegally take my kids to a playground (outside). The world seems upside down. Which should I do?

4:18 PM · Sep 26, 2020 from San Francisco, CA · Twitter for iPhone

Gummi Bear
@gummibear737

Professor Neil Ferguson Deep Dive

"Professor Lockdown", as he's nicknamed in some newspapers, is back in the public space and offering his scientific opinions on public health policy

As such, let's take a look at Neil and his most important work on pandemics - including C19

1

10:47 AM · Sep 17, 2020 · Twitter Web App

Who is Neil Ferguson? He was the Imperial College epidemiologist who presented the doomsday scenario to U.K. leaders. He predicted that 550K would die in UK and 2.2M in the US if lockdown measures were not undertaken promptly. ++ Lockdowns were never a part of any rational pandemic playbook. As I tweeted: "Lockdown is not a strategy, it's a panic move". Yet governments accepted his predictive modeling without criticism. How did Ferguson manage to convince the world to shut down? ++ In 2001, Ferguson made incredible predictions about the UK's foot and mouth disease outbreak. His research led to the culling of 11M sheep and cattle. Estimated 10B pound damage to UK economy. Other epidemiologists were critical but the govt followed his lead. ++ In 2002, there was a predicted outbreak of BSE (aka mad cow disease). Ferguson predicted between 50-55,000 people would possibly succumb. He later stated that up to 150K could die if certain conditions occurred. Ultimately, only 177 died. ++ In 2005, Ferguson predicted that 150M (up to 200M) could die of bird flu. Ultimately, <300 people died. Hmm, anybody seeing a trend? ++ When asked by The Guardian about this, he replied that "~40M died during the 1918 Spanish flu. There are 6x as many people now so you could scale it up to ~200M probably." ++ In 2009, Ferguson worked on swine flu predictions for the UK govt. His modeling suggested ~65K deaths. Ultimately, ~450. His team had estimated an IFR or 0.3-1.5% (Neil's best guess 0.4%) when it was later determined to be 0.026%. ++ In 2020, Ferguson prepared his COVID model for the UK. The UK was faced with a novel coronavirus, very little hard information, the potential of 550K deaths, and overwhelming of the NHS. They blinked and accepted Ferguson's predictions with little criticism. ++ @SunetraGupta (epidemiologist at Oxford) was livid that the UK accepted Ferguson's model with so little debate and has been a staunch critic throughout. She recently said that 'puritanical' criticism of idea of herd immunity has chilled legitimate scientific and policy debate. ++ Ferguson was later forced to resign from his position due to an affair he was continuing during UK's lockdown. He was infected with COVID at the time and knowingly exposed the woman and her husband & children. "Rules for thee but not for me." ++ Ferguson's modeling was built on >13 year old code. It's been described as SIMS without graphics. It was designed for flu and not COVID. He has been extremely reticent about releasing the code for examination. Those who have seen it have at best not given it any praise. ++ The UK public policy decisions of lockdown and severe mitigation were based on this modeling. The base assumptions for the policies have taken on an almost religious dogma. Any researcher or work that challenges the assumptions is attacked or ignored. ++ Respected physician scientist researchers have had their reputations attacked and their motives impugned. Research/papers have not been accepted or are held to higher standards bc of their potential to challenge the public health policies and the public's reception of them. ++ [Thread continues with more information - https://twitter.com/gummibear737/status/1306605767630491650]

Kyle Lamb
@kylamb8

Today the CDC acknowledged what has been known for several weeks: the flu has largely disappeared this flu season in the southern hemisphere. The CDC is spinning this as being the result of mitigation (masks, lockdowns, etc.). This thread will show that to be false.

-1-

nmary

is already known about this topic?

nza activity is currently low in the United States and globally.

is added by this report?

ving widespread adoption of community mitigation measures to reduce transmission of SARS-CoV-2, the virus
s COVID-19, the percentage of U.S. respiratory specimens submitted for influenza testing that tested positive
ased from >20% to 2.3% and has remained at historically low interseasonal levels (0.2% versus 1–2%). Data from
ern Hemisphere countries also indicate little influenza activity.

are the implications for public health practice?

entions aimed against SARS-CoV-2 transmission, plus influenza vaccination, could substantially reduce influenz
nce and impact in the 2020–21 Northern Hemisphere season. Some mitigation measures might have a role in
ing transmission in future influenza seasons.

10:37 PM · Sep 17, 2020 · Twitter Web App

First off, WHO has NEVER found lockdowns/mitigation to stop the flu. Here is their guidance from 2019 saying isolation/quarantine simply doesn't work ++ Let's be more blunt, a working group by WHO found in 2006 that there has never been any evidence of quarantine/isolation strategies stopping the spread of the flu ++ So what's the deal? I believe it's the virus itself that's impeding the flu. But to show that, I need to eliminate mitigation as a factor. More on that on a moment. But I have been calling this: they're banking on a light flu season to take credit. ++ The flu has been seen at historically low levels in the Southern hemisphere. Under 2% positive rate, in fact. Most years, WHO reports southern temperate regions to be between 10-30%. Not a single one has been at 10% this year. ++ So if it's not mitigation, why did the flu disappear then. As I said, I think it's the virus itself causing it to disappear. To illustrate this, let's use a normal flu season here in the U.S. ++ First, here are the last three years for confirmed Flu hospitalizations in the U.S. for calendar weeks 1-17. These are on a per 100,000 basis. You can see a steady, reliable downward trend in the yellow weeks 10-17. ++ Now let's look at Flu surveillance for cases and tests in U.S. clinical labs by the same weeks for the same three years. The highlighted columns are percent positive tests. Again, we see positive testing fall slowly and consistently. ++ Now that we have established our baseline for flu trends, let's go back to hospitalizations and add in this flu season (2019-20). Strangely (or not), flu hospitalizations fell off a cliff starting after week 10 (week ending March 7). ++ You might ask, well is it because of lockdowns? They didn't start until weeks 12-14 in most areas and wouldn't have changed the numbers that quickly. Was it no more flu testing? No. See what happens to the percent positive testing in the same timeframe. ++ Almost by magic, while testing did not really decrease much until at the very end, the percent of tests that were positive just completely vanished. Gone. Flu was nowhere to be found. ++ To me, since flu disappeared here in the U.S. when Covid-19 arrived AND disappeared in the southern hemisphere where Covid was already lurking when flu season began, this demonstrates circulation of Covid-19 is causing disruption in flu spread; NOT mitigation causing this. ++ Addendum #1: Folks are still trying to argue lockdowns helped and this is a terrible argument. I have receipts. Only 3 stay at home orders took effect at the very end of MMWR 12 (March 21); AK, CA & CO. Rest were 13/14 Lab results are when REPORTED to CDC (not collection) ++ So most results reported week 13 were likely from week 12 or possibly before. It would take an additional few weeks for mitigation to have any impact. At minimum, if lockdowns had ANY impact it wouldn't show until ~14 or 15. By then, its' very clear from the data flu was done.

TEAM REALITY

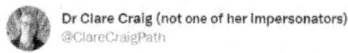
Dr Clare Craig (not one of her impersonators)
@ClareCraigPath

I had assumed non-COVID excess deaths in April were likely undiagnosed COVID. This coroner's audit suggests more likely lockdown deaths. Only 2/67 autopsies had COVID: thelancet.com/journals/lanpu...

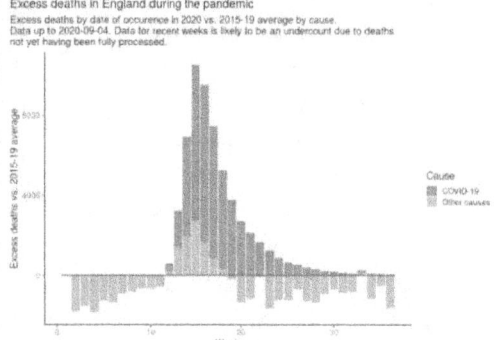

7:02 AM · Sep 27, 2020 · Twitter Web App

Emma Woodhouse
@EWoodhouse7

Why is the Gov of IL allowed to keep extending a 30-day emergency order? We are at 6 times.

Intent of law that allows disaster declarations is not for Gov to assume power indefinitely.

It's for him to be able to act swiftly in a true emergency. Thereafter, Leg branch needed.

12:52 PM · Sep 11, 2020 · Twitter for iPhone

Woke Zombie
@AWokeZombie

#NJ #COVID19 Saw this elsewhere, decided to recreate for NJ. This is for today using 30 day avg for Covid deaths from the DoD data.

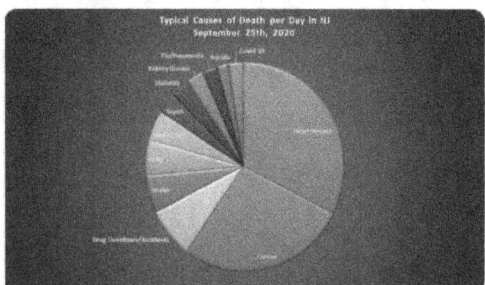

2:33 PM · Sep 25, 2020 · Twitter Web App

Jenin Younes
@Leftylockdowns1

even if we ever go back to "normal," I think I'll spend the rest of my life in fear that at any minute everything can be taken away

6:46 PM · Sep 14, 2020 · Twitter Web App

Emma Woodhouse
@EWoodhouse7

Since mid-March, more than 3x as many Chicago residents age 39 & under died by a gun (suicide or homicide) than died with/from COVID19.

Does not yet include gun deaths from this weekend.

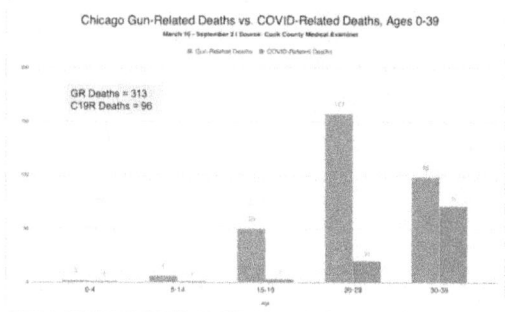

2:07 AM · Sep 7, 2020 · Twitter Web App

Gummi Bear
@gummibear737

Has there ever been a National Emergency in which our government and media have sought to instill fear and panic in the general public?

First rule in a crisis is don't panic

Stay calm, keep your head

Lockdown is not a strategy, it's a panic move

It's been 6 months...

1:11 PM · Sep 15, 2020 · Twitter for iPad

Eric
@The_OtherET

In case anyone had any doubt how criminally dishonest the media is regarding politicians' handling of #COVID19, the states with the governors with the 10 highest approval ratings is the blue curve. The rest of the US is the orange one.

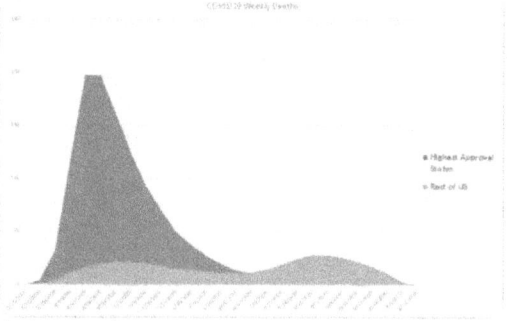

12:54 PM · Sep 15, 2020 · Twitter Web App

TEAM REALITY

Eric @The_OtherET

It's now been three weeks since the NYT article detailing how PCR testing set at too high a cycle threshold can lead to up to 90% of positive tests to be from people who are not infectious. Has ANY government, lab corp etc adjusted based on these findings? Genuine question

> **Don Wolt** @tlowdon · Aug 29, 2020
> Stunning new research finds C19 PCR tests are much too sensitive to be useful in assessing spread. "In 3 sets of testing data...compiled by officials in MA, NY & NV, up to 90% of people testing positive carried barely any virus." The implications are huge.
> nytimes.com/2020/08/29/hea...
> Show this thread

10:37 AM · Sep 20, 2020 · Twitter for iPhone

Ann Bauer @annbauerwriter

We've had 6 months to learn who's vulnerable to Covid. The elderly, poor, minority, chronically ill or worn by life. Yet we persist in protecting white, professional office workers while those with comorbidities ride public transit and wait on us at Walgreen's and Costco.

7:49 PM · Sep 27, 2020 · Twitter Web App

Bethany S. Mandel ✓ @bethanyshondark

Please stop captioning your photos with justifications about how you're six feet apart, were wearing masks, etc. You are not beholden to the Internet about how you live your life.

9:21 AM · Sep 29, 2020 · Twitter Web App

Zac Bissonnette @ZacBissonnette

People keep sending me weird covid conspiracy theory links. Just so everyone knows:

I don't think there's a conspiracy here.

I think there is an emotional overreaction that is completely unmoored from actual evidence—driven by hysteria and politics.

That's all.

11:54 PM · Sep 26, 2020 · Twitter for iPhone

Prof Francois Balloux ✓ @BallouxFrancois

I had never fully realised until now that the reason pandemic brought down so many empires and kingdoms in history, wasn't the death toll, but the fear, the sense of doom, the irrationality and the disunion they unleashed.

11:47 PM · Sep 9, 2020 · Twitter Web App

TEAM REALITY

OCTOBER 2020

Justin Hart
@justin_hart

Govt: "Those most at risk to COVID-19 are those with co-morbidities frequently stemming from obesity, lack of physical exercise, & low Vitamin D… so sit in your homes, do nothing for a few months, grow obese until we think you're ready to come out and face the virus!"

11:51 PM · Oct 20, 2020 · Twitter Web App

TEAM REALITY

Prof
@covidtweets

Remember how, when people were protesting lockdowns, many were suggesting that the protestors needed to waive their rights to healthcare? We need the lockdown proponents do the same - waive their rights to use "essential" services.

Here, I made it easy for them:

> **LOCKDOWN AGREEMENT**
>
> I, _____, believe another lockdown is necessary to contain the virus. As a result, I agree that (check only one):
>
> ☐ I will not be using any of my utilities, get food delivered, or travel for any reason. I will survive by myself for the duration of the lockdown.
>
> ☐ My life is more important than the people who serve me, so it is OK if some people die while preparing or delivering my food, getting me my groceries, etc.

11:55 PM · Oct 4, 2020 · Twitter Web App

Michael P Senger
@MichaelPSenger

Historians may come to refer to Xi Jinping as the first person to ever rule the world—albeit briefly.

By early April 2020, nearly the entire world was under some form of lockdown, per "the science"—"the science," of course, being the will of Xi Jinping, Chairman of the World.

12:36 AM · Oct 21, 2020 · Twitter Web App

Gummi Bear
@gummibear737

Yay, the Billionaire Report is out!

Despite COVID-19, there are more billionaires and their net worth has increase to $10+ trillion

But working class people and small business owners: stay home...just because

This is becoming very sinister

#GreatReset

3:18 PM · Oct 9, 2020 · Twitter for iPad

Jennifer Cabrera ✓
@jhaskinscabrera

A little over 3 weeks later, Florida is fine. In fact, hospitalizations yesterday hit a record low since the July peak before bouncing back up just a bit today (normal for Monday).

Case positivity rates have held steady.

Once again, Fauci's predictions of doom were wrong.

> **MSNBC** ✓ @MSNBC · Sep 29, 2020
> Dr. Fauci says Florida's full reopening of bars and restaurants is "very concerning." on.msnbc.com/33cSgCx

10:41 PM · Oct 19, 2020 · Twitter Web App

Virál Myãlgía MD, PhD
@contrarian4data

Vote for rational trade offs, self determination, and sending children back to school this year.

I'm a single issue voter this year.

12:07 AM · Oct 29, 2020 · Twitter for iPhone

Erich Hartmann
@erichhartmann

Let's be clear here:

A vocal, elite minority who can work from home indefinitely, without fear of losing their jobs, who can afford childcare or don't have kids, and have an almost religious faith in government control... they are Team Lockdown.

The 1% have done this to us.

10:39 AM · Oct 29, 2020 · Twitter Web App

Justin Hart
@justin_hart

Remember when we were trying to "save Easter"?

7:00 PM · Oct 27, 2020 · Twitter for iPhone

Real Developments
@pdubdev

Covid lockdowns are destroying global supply chains and will lead to famine and starvation.

These are Scenes from Nigeria food distribution.

Your virtue is their death.

twitter.com/YellowCube7/st...

7:15 PM · Oct 27, 2020 · Twitter for iPhone

My Fauci Deep Dive

America's most famous public health expert:
- Exalted deity to some
- Embodiment of evil to others

What's the truth?

It's complicated...

It starts slow, but boy does it get good

Please RT this first tweet if you learned something new and interesting

Any retrospective on Fauci has to start in the 1980's with the arrival of HIV/AIDS. A trained immunologist, he was named director of the National Institute for Allergens and Infections Disease (NAIAD) in 1984, and has held this position since (36 yrs). AIDS was his first test ++ Initially, Fauci was the target for the frustration of the Gay community which was being disproportionately affected by AIDS. Many were dying and treatments were not coming fast enough. Larry Kramer famously penned this critique of Fauci: https://villagevoice.com/2020/05/28/an-open-letter-to-dr-anthony-fauci/ ++ Ultimately Fauci agreed to fast-track treatments: "If I had a disease in which the result was that I would die no matter what, and the government was telling me, You can't try anything that might work under any circumstances, I'd be ramming down the doors, too" ++ With the arrival of Highly Active Anti-Retroviral Therapy (HAART), the tide in the battle against AIDS. It was no longer a death sentence, it became a manageable chronic illness. Today, Fauci is viewed as a hero by the HIV/AIDS community. ++ There's also plenty of criticism to accompany Fauci's management of the AIDS crisis. I lived through it as a child and it was quite scary at the time. Fauci is responsible for creating a lot of the panic during the 80's. It's almost like he really craves media attention... ++ Since then, Fauci has remained at the NAIAD through 6 presidents- his 4 predecessors served terms of 9, 7, 9, and 9 years. He's been front and center with SARS, Bird Flu, Swine Flu, MERS, and Ebola outbreaks. And he knows the political game...Here: sending his love to Hillary. ++ I remember a few years when Gates announced his "Decade of Vaccines". Since, there's been this strange intersection between The Gates Foundation, the NIAID and the Fred Hutchinson Cancer Research Center. The triumvirate's been very active during C19. ++ Not so bad yet, but then we get to the gain of function research (GoF). Here we start seeing a link between NIAID, Ralph Baric (UNC Chapel Hill), and Dr Shi Zhengli (Wuhan Lab) with Peter Daszak (Ecohealth) in the middle. Baric+Shi made a GoF virus. ++ Now, let's get to Covid-19. First issue, he used the IC model from Ferguson to convince Trump to lockdown. "The steep curve with 2.2 million deaths was not from Dr. Fauci, however, but from Neal Ferguson's team at Imperial College". ++ Next, together with NIH Director Francis Collins and Larry Corey from FredHutch, they created the US Covid-19 Strategy. They wrote this paper to detail their plan. They convinced Trump to mobilize govt, pharma and academia to go all in on a vaccine. ++ [The thread continues with information about funding NIAID received and other facts]

TEAM REALITY

 **Michael Tracey** @mtracey

Perhaps the most radical position one could have taken since March is to apply the same standards for COVID mitigation equally across the political spectrum -- because only about .0001% of the commentariat has actually done so despite the constant theatrical posturing and lying

1:41 PM · Oct 5, 2020 · Twitter Web App

 **Ann Bauer** @annbauerwriter

For 20 years I have told my children never to make decisions out of fear -- because that almost always leads to catastrophe.

I double down on that advice.

9:35 PM · Oct 25, 2020 · Twitter for Android

 **ClownBasket** @ClownBasket

Viruses mutate. Wouldn't it be terrible if Covid-19 mutated to become MORE deadly for kids - in the meantime Governors across the USA implement lockdowns preventing kids from getting this virus now when it has very mild impacts? Protect the vulnerable; open up society.

7:33 AM · Oct 23, 2020 · Twitter for iPhone

 **Real Developments** @pdubrlev

Above threshold of 32 no live virus found.

Most labs cranked up their machines to 40.

Total sham.

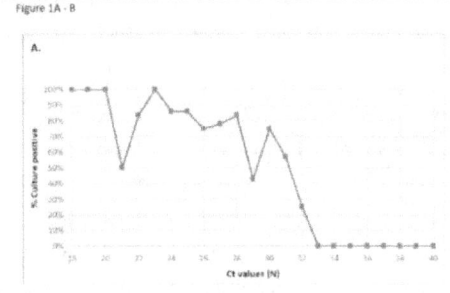

9:18 PM · Oct 26, 2020 · Twitter for iPhone

 **Stefan Baral** @sdbaral

#jobsplaining: employed people tell those laid off not to worry about economy
#houseplaining: those with housing tell people at risk for eviction not to worry
#backyardsplaining: those with backyards tell those without that closing parks is best

Let's not dismiss real concerns.

3:56 PM · Oct 11, 2020 · Twitter Web App

 Ian Miller @ianmSC · Oct 29, 2020

It is remarkable how seasonality has been completely ignored by experts & media

Hospitalizations per 1M in Arizona & Florida vs Illinois, Michigan and Wisconsin shows no matter how many masks are required and when, seasonality drives spread

When it's your turn, you get it

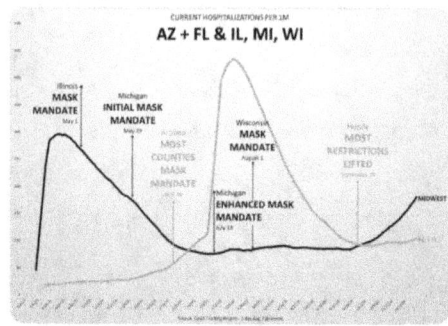

 **Prof** @covidtweets

When we think about the rough times of the past like the WW2, we mostly think about the heroic sacrifices people had to make during those times. I feel like future generations looking back at today's events will think about how dumb we all were.

10:41 PM · Oct 28, 2020 · Twitter Web App

 **Martin Kulldorff** @MartinKulldorff

Twitter has censored two tweets by scientist and White House Coronavirus Task Force member @SWAtlasHoover. After 300 years, the Age of Enlightenment has ended.

6:05 AM · Oct 18, 2020 · Twitter Web App

Prof
@covidtweets

There is an angle to all of this that I believe is being missed. Lockdown governors probably did not save lives, despite many claiming that they did. They either caused more deaths or at minimum moved deaths from one demographic to another. (1/x)

9:05 AM · Oct 3, 2020 · Twitter Web App

Why did they cause more deaths? Two reasons, one of which is obvious and the other one is subtle. The obvious one is that lockdowns meant higher SES staying home while lower SES "essential workers", who are generally in poorer health on average, going to work more. ++ So, while people thought they were eradicating the disease by staying home, they forced more low-wage workers go to work to feed them because now food, groceries, etc. had to be delivered. This created higher risk for people who are in generally poorer health. ++ Besides, those people went to work without other preventive measures being taken, because the panacea was lockdown. If masks work, they didn't even have that. Because we were already doing something, we did not focus on how to stay open and also stay safe. ++ The subtle reason is that the haphazard closure of all the businesses created an adverse reaction in a large portion of the population, especially after the initial fear subdued. People who would be on board with reasonable measures now became resistant to anything. ++ My theory is that, instead of lockdowns, which do more harm than good, if we had focused on reasonable measures and allowed the life to go on, most people would follow those measures, and more lives would be saved. Instead, we panicked and chose the most destructive path. ++ So next time someone blames someone for not wearing a mask, not distancing, etc., say this is largely lockdown governors' fault. The thesis created its anti-thesis. A moderate approach, on the other hand, would be more effective and more people would be on board.

John Ziegler
@Zigmanfreud

Of all the inherent contradictions in the news media/"expert" narrative on COVID, perhaps most remarkable is they somehow claim BOTH that immunity may not be be real or lasting AND that a vaccine is our only hope at getting through this.

They do know how a vaccine works, right?!

2:02 PM · Oct 15, 2020 · Twitter for iPhone

Tracy Hoeg, MD, PhD
@TracyBethHoeg

0.17% positivity rate in NY schools so far. Why is this surprising? It is not surprising if you have been following the worldwide epi data showing kids in k-12 are NOT major viral vectors."Surprising Results in Initial Virus Testing in N.Y.C. Schools"
nytimes.com/2020/10/19/nyr…

5:24 PM · Oct 19, 2020 · Twitter Web App

Justin Hart
@justin_hart

We've lost our everlovin' minds.

2:29 AM · Oct 24, 2020 · Twitter for iPhone

TEAM REALITY

Kyle Lamb
@kylamb8

98%.

What if I told you that confirmed flu surveillance across the world has dropped year over year by 98% since April?

This is the deep dive story with data, graphs and charts showing how the world's most consistent nemesis has (almost) completely vanished.

-1-

			Southern Hemisphere	
19 Flu	2020 Flu	+/-	2019%	2020%
2,169	1,900	-12.40	9.06%	7.39%
34,059	2,156	-93.67	12.99%	1.04%
32,741	374	-98.86	13.59%	0.23%

3:31 PM · Oct 16, 2020 · Twitter Web App

We began hearing reports back in May the flu was nowhere to be seen in the Southern hemisphere, which was the start of when cases usually peak. I had theorized as early as June and July that perhaps Covid was keeping the flu away. ++ Though just a theory I floated at the time, probably earlier and more consistently than most, it's one that now shares some scientific backing. Check out this study published Sep. 4 in The Lancet that corroborates the theory with some precedent. ++ Of course, there are other theories as to the disappearing act. One is sinister. Some say the flu never left, but the cases are showing up as Covid. I'm not here to advance that but it's out there. Others say masks & mitigation stopped it. I will show that to be silly. ++ Before we go further, here is the bottom line: using the WHO FluNet database from the Global Influenza Surveillance and Response System (GISRS), I have aggregated all flu cases by week from 2019 by WHO transmission zone (there are 18 of them) and also for 2020. ++ For WHO surveillance, weeks begin with a Monday and end on a Sunday. When splitting 14 zones into the Northern Hemisphere, 4 into the Southern Hemisphere, here is what we get for a comparison for the first 8 weeks, since week 10 and since week 15. 97.9% decrease since Wk. 15 ++ Let's start where this all began: China. Of course, the numbers there should be taken with a grain of salt, but the pattern will be unmistakable. Here is the WHO Eastern Asia flu transmission zone by week for 2019 compared to 2020. See how flu dives once Covid arrives. ++ This is no isolated case. Let's bring in three other Northern temperate flu zones (N. America - U.S. & Canada, Northern Europe and South/West Europe). As soon as Covid-19 cases show up, see how flu cases and % positive take a nosedive in all four zones. ++ Let's go one better and show a visual. Using a log scale, here is what 2019 cases (green), 2020 cases (red) and Covid-19 cases (yellow) look by week. See 2020 flu cases deviate from 2019 as soon as Covid intersects. The bottom is cropped because of literal zero cases. ++ It's important to remember a mere fraction of all flu cases are ever confirmed. And within those, a lower total is monitored within GISRS reporting. Still, we are about to see a shocking pattern of consistency. Nevermind that 98% of these surveillance cases went away. ++ Here are four more flu zones in the Northern tropic/sub-tropic regions. We have Central America/Caribbean, the Middle East, Southern and Southeastern Asia zones. Slightly different timing, same patterns. Flu goes away when Covid says 'hey' ++ Here, we can see the data showing how once Covid ascends, flu cases are immediately suppressed. It's important to see that testing does not really decline. In fact, since week 15 globally, flu tests are actually up over 2019. ++ In the Southern hemisphere, where the trend was first noticeable by experts, we confirm the flu never really got going. See the modest number of cases by week and zone in both South America regions, Southern Africa and Oceania (primarily Australia). ++ Now let's see those four Southern hemisphere zones in graph form. They follow a different timeline than Northern hemisphere but same result: once Covid-19 cases ascend, the 2020 flu cases depart from the 2019 trend. ++ Some argue masks, distancing & lockdowns were responsible. However, most zones began going down before these mitigation measures were in place. And to argue mitigation stopped the flu defies decades of scientific research and simultaneously destroys the idea it stops Covid ++ Some will still argue the point just because. Let's look at Japan, a place that is credited with disciplined mask wearing even during the flu... which didn't stop WHO from categorizing them as having "widespread outbreak" the past five flu seasons. ++ Let's look at the past four flu seasons for Japan, weeks 1-16. Compare them to 2020 and see how the cases fell off a cliff when Covid arrived and it happened before the first mitigation measure -- school closures -- even began. ++ Bottom line: confirmed flu cases were down only about 0.8% globally the first 8 weeks of the year. But as soon as Covid-19 began spreading, flu cases stopped. They're down YoY 69.4% since week 10 and 97.9% since week 15. ++ So what does this mean? Some will say they've replaced flu cases with the flu. I can't rule anything out but for now I'll stick with my theory from the summer that Covid-19 has pushed out the flu. This suggests fear mongering over a winter "twindemic" is misplaced. ++ Final addendum for the 'mitigation cured the flu' crowd... if that is true, you realize it means mitigation stopped flu dead in the middle of established outbreaks and allowed Covid to explode on to the scene without getting in its way. Is that really the hill you want to die on?

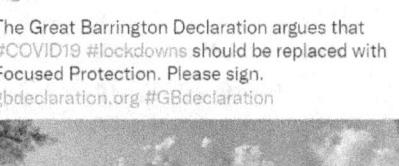

Masks Masquerading as ScienceTM.

I spent a good part of my career engineering emulsion PCR and Nebulization equipment for DNA and RNA shearing. I have a very different view on this topic and have mostly stayed out of the debate as it's now cultish.

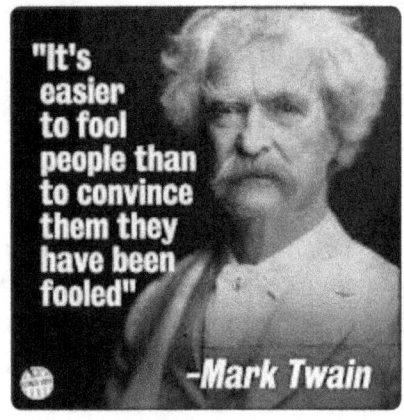

12:00 PM · Oct 24, 2020 · Twitter Web App

My largest concern over mask studies is most of them fail to include live virus in the study and measure the infectivity of the particles on the inside and outside of the masks, the fomite formation, and the microbiome risks associated with non-sterile technique mask use. ++ The focus as been myopic ironically. It is focused on the reduction of large droplets while failing to consider if the larger droplets are getting nebulized into smaller ones. ++ If you study the field you will see viral nebulization is a technique used to get higher infectivity of viruses. This is used with ventilator patients suffering from pseudomonas aeruginosa infections. Nebulize bacteriaphage to get deep into the lungs and kill bacteria. ++ In fact products exist on the marketplace to nebulize medicines because aerosols get deeper into the lungs than large droplets. ++ ...

++ Raises the question.. Are the masks stopping large droplets or nebulizing them into Aerosols? Aerosols are smaller than 5um and more infective. They are more monodispersed and less likely to precipitate out of the air with humidity. Why dont the mask studies use live virus? ++ I am not saying masks do or do not work for all settings. I've not seen compelling data that proves they do work in public settings. Ive seen alot of correlative noise but in medicine we first prove they work before demanding everyone use them. We seem to have flipped that around ++ I'd love to see a microbiome study performed on the masks that cycle from dirty pocket to dashboard multiple times a day. They certainly get moist which would imply bacterial growth in 18 hours. Imagine a world where everyone strapped a dirty sock on their face all day long. ++ Id also love to see a world where I could explain to children why we take them off when we sit down without having to appeal to our obedience to illogical tyrants. ++ Let's aside our science for a minute and assume they do work. For most of you this wont be hard. Its is no longer debated that our best protection from C19 has been the populations ancestral immunity to common cold coronaviruses like OC43, HKU1, NL63, 229E etc... + ... + Our ancestors paid it forward by building a vibrant free market economy that took these slight infections risks while building immunity. Our generation is too selfish. They want everyone else to hide to help them and have no time preference for the future. C19 is here. ++ If you believe masks work.. You should be worried about an immune forest fire you are creating by trying to centrally plan the human immune system. This is hubris IMHO. Boom and Bust cycles will become the norm. If you believe the science is unsettled, jurisdictionaly experiment. ++ ...

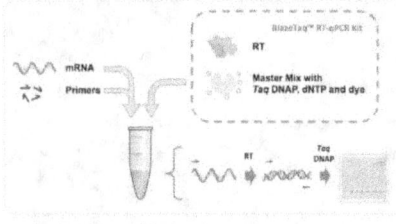

Kevin McKernan
@Kevin_McKernan

The Live-Dead qRT-PCR problem, the testing industrial complex and its impact on society.

I never thought the work I did for the human genome project would be weaponized to lock down society. We are now ruled by qPCR right and the transparency on the process is shameful.

9:23 PM · Oct 26, 2020 · Twitter Web App

I hope to convey the need to push for public transparency on the CQ values being utilized to count "case" #'s. I also hope to convey the need for home testing and how this regains our medical privacy. It is clearly required for any personalized medicine dream to materialize. ++ Why listen to me? I've done a lot of PCR (25 yrs). Below is factory floor our R&D team designed, programed, and maintained at The Whitehead Institute/MIT Center for Genome Research. It processed 40M Sanger sequencing reactions/year (a more complicated version of linear PCR) ++ What is the live dead PCR problem? PCR doesn't count infectious virus. It counts RNA molecules which can be from live or dead virus. Initial infections contain lots of live virus but as you clear the virus the Live-dead ratio shifts more towards the dead. ++ ... ++ Jaafar et al. showed that Cqs after 33 were mostly dead. Most testing facilities report a positive result if under CQ 40 (but this should be verified in your jurisdiction). ++ Cqs are an inverse Log2 scale. Each Cq digit is a factor of 2 so a Cq of 10 is twice as much RNA as a CQ of 11. There are ~3.3 Cqs for every factor of 10. One would use 2^7 or 128 fold to estimate how much difference there is for 7 Cq offset from 33 to 40. ++ LODs .Limit of Detection. Take RNA, perform a serial dilution in qPCR and measure how low you can go. Most tests crap out 50 copies of SARs RNA at around 37 cycles. Calling positives past your Limit of detection is shady. I think there is a lot of this going on. ++ This can create False positives that can transform a pandemic into a "case-ademic". There are a few papers circulating that demonstrate this effect on other outbreaks that have occurred in the past. Please post them to this thread if you have them at your fingertips. ++ ... ++ Why Live-Dead matters. The majority of the time a patient is qPCR positive (<40Cq) is the tail end of the disease where the virus is shedding and more dead than alive RNA. I call this the lower infectivity long tail. People detected in this window are usually non-infectious ++ These poor folks are quarantined, track & traced, medical privacy violated, constitution nullified and shamed from society. But this creates a lot more testing volume! I doubt there is motivation to tighten the dial, despite there being cheap old-school tools which address this. ++ Everyone caught in the long tail create a chain reaction of contact tracing testing and when these people are young and have less risk than the flu... Great Barrington Declaration... @aier ++

Kevin McKernan
@Kevin_McKernan

You see inflation in the price of goods through the artery in which the money is being injected. Don't believe me.. qPCR test are usually under $10. CLIA regs might inflate that cost to $50. Money Printer go BRRRR ...The rapid COVID tests in Mass are $140.

10:08 PM · Oct 25, 2020 · Twitter Web App

TEAM REALITY

Vanessa @vlai42 · Oct 18, 2020

My family plans to trick or treat and have fun. These recs are silly. No kid wants to do a zoom costume contest. Labeling a risk moderate when you are six feet apart, masked and outside is silly. This is why people have tuned out recommendations. They simply do not match reality.

Angry Cardiologist
@AngryCardio

I am calling for a "Tournament of COVID FUD" (fear, uncertainty, & doubt).

This can include concepts (eg COVID myocarditis), actions (spraying streets with disinfectants), groups or individuals.

Send me your nominations & I will seed & run the single-elimination tournament.

10:23 AM · Oct 5, 2020 · Twitter Web App

Vinay Prasad, MD MPH
@VPrasadMDMPH

Glad we are willing to do the important things to stop COVID19 like removing swings from outdoor playgrounds

1:36 PM · Oct 18, 2020 · Twitter for Android

Angry Cardiologist @AngryCardio · Oct 23, 2020

Woohoo!

The #COVIDFUD Tournament has a champion!

Congratulations to Eric Feigl-Ding @DrEricDing for a resounding victory!

Eric is the Champion of COVID FUD!

7:01 AM · Oct 21, 2020 · Twitter for iPhone

80

TEAM REALITY

Prof
@covidtweets

Upon careful deliberation, I hereby announce that I am leaving team reality and joining team apocalypse. Below I list my reasons to do so, which, after you all read, should convince you that it is for the best to leave team reality. (1/x)

1:29 AM · Oct 2, 2020 · Twitter Web App

First, I will no longer need to scrutinize studies, analyze data, question the messaging, and request evidence of whether a measure against COVID actually works. I will just believe in what I am told. Much less cognitive effort. ++ Second, when confronted by a member of team reality, I will not have to actually share evidence of what I claim to be true. I will just throw a few insults and then block. That will save a lot of time! ++ Third, I will only focus on one outcome, which is eradicating COVID and eliminating only COVID deaths. Much more straightforward! I will not have to think about other harm being done in the process, as that harm is happening unseen and will take time to be trendy anyway! ++ If this is not enough justification to convince you that team apocalypse is the obvious choice, I don't know what is and you all are morons! Have fun with your data and your search for evidence! You all are blocked!

Prof
@covidtweets

When it is all said and done, we will realize that kids and young adults have suffered the biggest harm from this pandemic. The prevailing response of closing schools, businesses, etc. and restricting social life is devastating to everyone, but especially to the young. (Thread/1)

10:35 PM · Oct 23, 2020 · Twitter Web App

First, the obvious: The younger are losing months, if not years of schooling. They are falling behind in life. They are missing out on their childhood as playgrounds, theme parks, playdates, etc. are denied to them. They are losing sports. Their mental health is suffering. ++ The older are missing out on their high school/college experiences. Their school is being reduced to looking at a screen all day. They are losing friendships. They are losing proms, graduations. They are losing sports. Their mental health is suffering. ++ Next, the not so obvious: They are losing future relationships, which are essential to finding jobs, partners, a meaningful life. An internship, a job fair, a prom, a convention... These are important milestones for careers and marriages alike. They are not happening. ++ As governments are in a contest of which one is the most stupid, the businesses are shutting down, unemployment is spiking, etc., creating a massive bill as spending is up and revenue is down. This bill is accumulating, and it will be the young who is going to pay for it. ++ These harms are not on CNN because they are not trendy yet. They are hard to quantify. But tens of millions of people, while not losing their lives, are losing their chances of a meaningful life, which is just as important. The irony is, for a virus nearly harmless to them.

TEAM REALITY

Hold2
@Hold2LLC

This has been my main issue with the mask mandate argument from the beginning. People have become so singularly convinced that masks "work" without proof, which has led to 2 things:

1) Gov't leaders thinks that's all they have to do
2) People get distracted from real causes

/1

> **Wes Pegden** @WesPegden · Oct 23, 2020
> Studies like this overstate our quantitative understanding of the benefits of mask mandates.
>
> This does a disservice both by giving the impression that the issue doesn't warrant further investigation, and by drawing attention from the need to improve our response in other ways.1/ twitter.com/ScottGottliebM...
> Show this thread

9:06 AM · Oct 24, 2020 · Twitter for iPhone

Emma Woodhouse
@EWoodhouse7

I'm told that as of yesterday, University of Illinois athletes in at least one winter sport are now being tested for COVID twice a DAY, seven days a week.

Am I alone in thinking that's a monumental waste of time, money, resources, & sanity?

3:03 PM · Oct 26, 2020 · Twitter Web App

Gummi Bear
@gummibear737

California's Holiday Gathering Covid Rules
-Only three families together at once
-Guest tracing
-Gatherings outside only, internal bathroom ok if routinely sanitized
-Everyone must sit 6 feet apart
-Food in single serve containers only
-Only 2 hours allowed

Lost their minds...

7:29 AM · Oct 21, 2020 · Twitter Web App

Eric
@The_OtherET

Florida ended its state-level #COVID19 restrictions four weeks ago today.

At that point, it had 2,155 hospitalized patients.

Today, 10/23, four weeks later, it has 2,087.

Reminder: the whole reason restrictions were put in place to begin with were to not overwhelm hospitals.

6:44 PM · Oct 23, 2020 · Twitter Web App

Gummi Bear
@gummibear737

In my humble opinion

Based on CDC data

And recent decisions by Nordic countries

Mandating the vaccination of children is evil

1:44 AM · Oct 8, 2021 · Twitter for iPad

Emma Woodhouse
@EWoodhouse7

For the record, if you believe the decline in flu is attributable to masks, you're gonna have to explain how your piece of cloth selectively deters flu but not COVID or rhinovirus.

4:41 PM · Oct 17, 2020 · Twitter Web App

Bethany S. Mandel
@bethanyshondark

"What lockdowns? I don't see any lockdowns!" They gaslight from the house they've spent seven months not leaving, applauding themselves for ordering Instacart and doordash. WHAT COURAGE. WHAT SACRIFICE. (As they pay poor people to do their shopping for them.)

11:05 PM · Oct 24, 2020 · Twitter for iPhone

Bethany S. Mandel
@bethanyshondark

The only inkling they have of the small business collapse is when they notice how there aren't as many restaurants available to order from as the weeks and months go by. No matter, must be a Doordash glitch.

11:22 PM · Oct 24, 2020 · Twitter for iPhone

Jennifer Cabrera
@jhaskinscabrera

I am SERIOUSLY worried about children. We are completely warping their view of the world, and that will stick with them.

I'm also very worried about people in long-term care, who are losing their last bits of time with their families.

All of this needs to stop. Now.

> **Scott W. Atlas** @ScottWAtlas · Oct 20, 2020
> Children are being literally destroyed. But, hey, better keep testing them, finding asymptomatics, confining low risk young people, test the college sewage system, limit in-person classes - "must" find those cases! And btw, nah, that's not lockdown... bitly.ws/albv

11:03 PM · Oct 20, 2020 · Twitter Web App

TEAM REALITY

NOVEMBER 2020

M_P
@Reroot_Flyover

There is no Public Health without public trust.

8:06 AM · Nov 22, 2020 · Twitter Web App

Nick Foy
@TheNickFoy

Staying home saves lives and everyone should just stay home.

Except for the DoorDash lady who delivers my dinner that was cooked by people who work at the restaurant.

And the grocery delivery driver who brings me the groceries that were stocked by the grocery worker.

1:09 PM · Nov 15, 2020 · Twitter for Android

And the Amazon worker who packed the stuff I bought. And the UPS driver who delivered it. And the brewer who made the beer I'm drinking. And the trucker who brought the beer to the store. And the football players who are playing in the game I'm watching. ++ And the linemen who fixed my power lines so my power came back on. And the plumber who fixed the sink that was backed up. And the cast & crew on this new Netflix show I'm watching. And the people who work at the water treatment plant making sure my water is clean. ++ And the people in the factory who made these shoes I just bought. And the longshoremen who picked the container that had my shoes off of the ship and put it on a train that ended up on a truck that made it to a warehouse and then somehow magically appeared on my doorstep. ++ And the daycare workers who are watching my kids so I can finally get some work done. When I said "staying home saves lives" it turns out I actually meant "me staying home saves my life."

Real Developments
@pdubdev

COVID is seasonal. Masks and lockdowns make no difference.

7 Midwestern states all move in perfect unison.

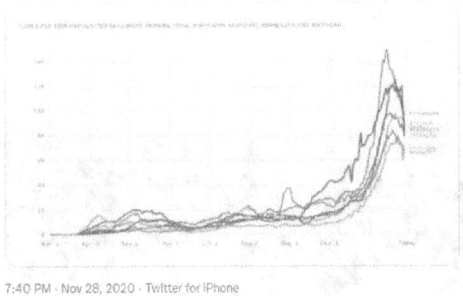

7:40 PM · Nov 28, 2020 · Twitter for iPhone

Don Wolt
@t1owdon

Study of Marine recruits finds CoV2 spread occurs despite strict quarantine & enforcement of pub health measures. This has major implications. If transmission occurs in this controlled environment, NPIs imposed upon the public would be even less effective.
nejm.org/doi/full/10.10...

2:08 PM · Nov 12, 2020 · Twitter Web App

John Ziegler ✓
@Zigmanfreud

I REALLY hope the teachers who have this week off are enjoying their vacation after their grueling first couple of months of pretending to teach via zoom & standing silently by as their unions & puppet politicians engaged in overt child abuse by keeping schools needlessly closed!

12:45 PM · Nov 23, 2020 · Twitter for iPhone

Federico Andres Lois
@federicolois

Today was the day!! The whole 'Lockdown Club' from the first wave has finally pushed out the bad boy out of the top places in the ranking. Without no fuzz Sweden just moved to the #18 with France taking the 'Precautionary Tale' place.

[table image]

↳ **Federico Andres Lois** @federicolois · Nov 5, 2020
Today is likely to be the day...
Show this thread

1:53 PM · Nov 5, 2020 · Twitter Web App

Victoria Fox
@drvictoriafox

COVID has really highlighted for me that science has a communication problem. I really hope we find our way back to inspiring public support with stories of the amazing life changing contributions science makes to the world, vs arrogantly berating them for not being "followers".

6:50 PM · Nov 15, 2020 · Twitter for iPad

TEAM REALITY

Stefan Baral @sdbaral

I'm not sure who needs to hear this, but arbitrary public health interventions undermine the public's trust in public health and burns their energy.

We need the public to trust public health.

And we need them to have the energy to follow through on high-yield recommendations.

9:53 PM · Nov 2, 2020 · Twitter Web App

John Ziegler @Zignanfreud

This may be my last tweet for a while & maybe ever.

Tonight, me & 15 relatives enjoyed Thanksgiving dinner, in violation of our King's guidelines.

It was indoors, 0 masks, 0 social distancing, & it was over 2 hrs.

I'm assuming the carnage which now ensues will be catastrophic.

9:08 PM · Nov 26, 2020 · Twitter for iPhone

Justin Hart @justin_hart

We've lost our ever-lovin' minds
h/t @drdavidsamadi

5:07 PM · Nov 25, 2020 · Twitter Web App

AJ Kay @AJKayWriter

I'll make a deal w/ all those saying that anti-lockdown folks should sign their rights to COVID-related healthcare away.

I'm game - as long as you sign away your right to use the schools we re-open or the industries we rebuild and refuse all jobs we create...

4:37 PM · Nov 28, 2020 · Twitter Web App

Tracy Hoeg, MD, PhD @TracyBethHoeg

Celebrating our sons getting A's instead of F's now that we moved them from a public (100% Zoom) to a private (in-person) school. Teachers' unions and others working to keep elem schools closed have no idea what kind of negative impact they are having on the kids of California.

6:46 PM · Nov 2, 2020 · Twitter Web App

Martin Kulldorff @MartinKulldorff

"Lockdowns now present the greatest threat to population health .. plummeting vaccination rates .. not getting diabetes treatment .. not attending for cancer screening, cardiovascular disease outcomes are worsening, and .. a huge strain on mental health."
thelancet.com/journals/lanre...

10:02 AM · Nov 25, 2020 · Twitter Web App

Martin Kulldorff @MartinKulldorff

Without public trust, public health is doomed. Despite heroic efforts by the public, #lockdowns and contact tracing have proved ineffective for #COVID19 and destructive for other health outcomes. To slowly regain trust, more public health scientists/officials need to speak out.

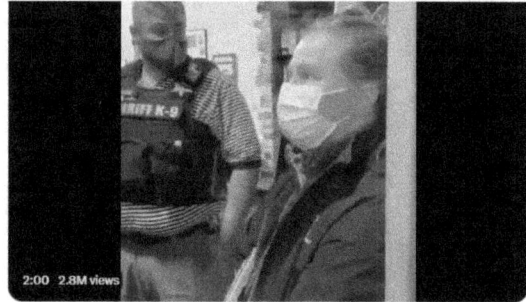

> Justin Hart @justin_hart · Nov 21, 2020
> Business owners in Buffalo, NY demand "health inspector" leave private property. "Go get a warrant."
>
> People have had enough of the #COVID19 tyranny curbing their life, liberty and the pursuit of happiness.

7:42 AM · Nov 22, 2020 · Twitter Web App

Prof Francois Balloux @BallouxFrancois

It feels at times as if the enlightenment age never happened. In many ways, our response to #COVID19 seems to me not that different from our ancestors' reaction to the Black Death, with its toxic mix of superstition, millenarianism, dogma, fear, anger and ostracism.
2/

3:13 AM · Nov 21, 2020 · Twitter Web App

TEAM REALITY

Pajamas It Is
@HeckofaLiberal

At some point we need to take a long hard look at what we are destroying based on misplaced fear.

4:37 PM · Nov 12, 2020 · Twitter for iPhone

Vinay Prasad, MD MPH
@VPrasadMDMPH

The best person to decide what's true and false in science are the employees of a multi billion dollar company whose primary source of revenue is advertisements

9:11 PM · Nov 27, 2020 · Twitter Web App

Eli Klein
@TheEliKlein

Keeping schools closed and limited is causing long-term damage to 10s of millions of children and untold future devastation on all aspects of life in the US

We don't know the extent of damage this tragic experiment caused, but it's the worst self-inflicted wound in US history

11:08 AM · Nov 21, 2020 from Manhattan, NY · Twitter for iPhone

Prof Francois Balloux
@BallouxFrancois

I wonder at times why so many in society seem to believe that if we run enough tests, we would automatically stop the transmission of a highly infectious pathogen with a short incubation period.

7:27 PM · Nov 8, 2020 · Twitter Web App

Wes Pegden
@WesPegden

It seems like the message has moved from

"close the restaurants so we can open schools"
to
"close the restaurants so we don't look like hypocrites for closing schools".

How about:

Open the schools because they are relatively safe, and it does irreparable harm to close them.

11:54 AM · Nov 12, 2020 · Twitter Web App

Mark Changizi
@MarkChangizi

If they feel justified to take your job, shut down your business, regulate what happens in your home, tell you what to wear, and force you to take medicines, then they will feel justified to do much MUCH more to you.

10:15 AM · Nov 27, 2020 · Twitter Web App

Vanessa
@vlal42

I am not an anti-vaxxer, I do not think covid is a hoax, I care deeply abt the elderly, I believe in science and I think our treatment of the kids during this pandemic is criminal.

8:42 AM · Nov 18, 2020 · Twitter for iPhone

Vinay Prasad, MD MPH
@VPrasadMDMPH

Washington state gov. now requiring kids to wear cloth masks for outdoor sporting events at all times

PS this is the weather

Recommendations like this lack evidence, lack basic, common sense and create backlash

governor.wa.gov/sites/default/...

2:54 PM · Nov 16, 2020 · Twitter Web App

Martin Kulldorff
@MartinKulldorff

In a large Danish randomized trial, 1.8% of mask wearers and 2.1% non-mask wearers were infected with SARS-CoV-2, p=0.38. The 95% CI is compatible with a 46% reduction to a 23% increase in infection. So, either modest or no benefit to the mask wearer.
acpjournals.org/doi/10.7326/M2...

10:05 AM · Nov 18, 2020 · Twitter Web App

Bethany S. Mandel
@bethanyshondark

Instead of being in classrooms, distanced and with masks on, NYC high school students will be mingling in their communities unsupervised while their parents work. They won't be learning, but they will be spreading. Closing schools isn't about virus mitigation.

8:06 AM · Nov 19, 2020 · Twitter for iPhone

TEAM REALITY

AJ Kay
@AJKayWriter

Keep the "Hosptializations from" vs. "Hospitalizations with" COVID data in mind when you see this headline circulating Twitter tomorrow: (Thread)

1/9

11:57 PM · Nov 16, 2020 · Twitter Web App

IA and ND are two states that delineate between hospitalizations "with" & "from" COVID. If, like most states, they did not differentiate, each state's COVID hospitalizations would be overstated by 38% and 20%, respectively. ++ Overstatements matter b/c hospitalization data is used in two ways: 1. as a measure of disease severity i.e. "the proportion sick enough with C19 to require hospital care" 2. to measure the threat of overwhelming hospitals. Overstatements inflate the perception of both... ++ A common belief is that overreacting is harmless, and even in line with the precautionary principle. However, overreacting is not, in fact, harmless and there can be immediately counterproductive consequences... ++ For example, the perception of impending crisis can drive communities to close in-person schools--as a "precaution"--despite evidence that schools aren't sites of significant transmission. However, closing schools actually results in reduced HC capacity & fewer staffed beds ++ In fact, that potentiality was the focus of summer modeling studies--exploring the impact of closing schools on C19 outcomes, given resultant healthcare workforce depletion. This problem is gaining attention right now... ++ Today, Judy Rich, the CEO of a hospital system in Tucson, AZ implored local school superintendents to offer in-person school options due to a nursing and med tech shortage caused by parents dropping out of the healthcare workforce... ++ Fewer healthcare providers = fewer staffed beds = decrease in capacity to care for COVID & non-COVID patients. And that's just one of many examples of why it's critical that the data disseminated to the public be accurate, contextual, & transparent. There are many more. ++ Our responses must be well-reasoned & proportional & that cannot happen without accurate data to make those decisions. Knowing how many are in the hospital for COVID (vs. with COVID) is crucial in developing responses that actually mitigate harms, not exacerbate them.

John Ziegler
@Zigmanfreud

History of masks/COVID:

March: Fauci mocks

April: In panic they become virtue-signal

May: Liberals realize their base WANTS mandates

Summer:They don't work anywhere (except where 1st wave is over)

Fall:3rd wave gets blamed on not enough mask wearing

Solution? More mandates!

10:45 PM · Nov 18, 2020 · Twitter for iPhone

Angry Cardiologist
@AngryCardio

If Public Health recommendations don't align with how normal people act and behave, who is to blame?

The people, of course!

What a stupid question.

10:07 PM · Nov 12, 2020 · Twitter for iPhone

Phil Kerpen
@kerpen

It is remarkable that anti-schoolers still exist.

How can anyone hate children and science that much?

7:33 PM · Nov 20, 2020 · Twitter for Android

TEAM REALITY

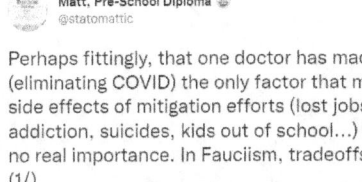

Matt, Pre-School Diploma
@statomattic

Perhaps fittingly, that one doctor has made one metric (eliminating COVID) the only factor that matters. Any side effects of mitigation efforts (lost jobs, drug addiction, suicides, kids out of school...) are ascribed no real importance. In Fauciism, tradeoffs barely exist. (1/)

> **Jennifer Cabrera** @jhaskinscabrera · Nov 14, 2020
> In my opinion, one of the biggest mistakes the United States has made regarding COVID-19 was anointing one virologist (Dr. Anthony Fauci) as the "expert" and ignoring all opposing views.
>
> 8/13
> Show this thread

7:47 PM · Nov 14, 2020 · Twitter for Android

This myopia frees up Fauci to recommend increasingly wacky solutions (hard lockdowns, mask mandates, no school...) b/c any negatives are irrelevant. Lockdowns might sort of work, so do them! Masks might help. Mandate them for everyone everywhere! ++ Couple this with the phenomenon @jhaskinscabrera mentions about all contrary views being shunned as anti-consensus and anti-science, and you get our truly toxic real environment of this year. A virus has been weaponized and politicized, its reported cases fetishized. ++ Risky mitigations with unproven (at best) positive value are accepted as normal, as necessary, as "following The Science™." All alternative viewpoints are dismissed as denialism, as a combination of selfish and stupid. In reality, they are none of these things. ++ Politicians are praised for ever-harder damaging lockdowns, for keeping kids out of school, for shaming Thanksgiving celebrations... Their "science" is assumed unassailable, and promoted by media propaganda. Meanwhile, in the real world... ++ A generation of kids goes without school, their development permanently damaged, Millions without jobs. Millions with unnecessary physical and mental health problems, citizens taught to fear and turn against each other. But none of that matters because only one metric does. ++ This is not science. It is politics. It is madness. It is evil. And it needs to end. That is all.

Prof Francois Balloux
@BallouxFrancois

I went through the COVID-19 news. I summarised it below, so that you can all have a lovely time doing something meaningful instead, rather than getting upset.

"There are other COVID-19 faiths that claim to be true but I know that the faith I'm following is the one true faith."

1:36 PM · Nov 22, 2020 · Twitter Web App

Prof
@covidtweets

I have been wondering if this could be the main difference between team R and A. I feel like many of us in team R have been following this thing since it's early days in China, Iran, and Italy, and knew what it was before the media freaked out and took everyone with them.

> **Jennifer Cabrera** @jhaskinscabrera · Nov 1, 2020
> Replying to @insahmity and @OBusybody
> Nearly everyone on Team Reality started by tracking the data themselves. It cures fear quickly.

10:41 PM · Nov 1, 2020 · Twitter for Android

Hold2
@Hold2LLC

It's bad enough that CNN would clown itself with a headline like this, but here's what happens next:

- Northern states will hit natural peak and decline (just like always - look at WI)
- FL will peak later
- When northern states drop below FL, media will target FL fully (again)

> **Hold2** @Hold2LLC · Nov 23, 2020
> Replying to @CNN
> What did Illinois, Michigan, Wisconsin, New Mexico, and New Jersey do since late September?
>
> Did you consider other factors, such as seasonal respiratory illness trends before trying to blame FL's numbers on reopening while the rest of the country has increased even more?
>
>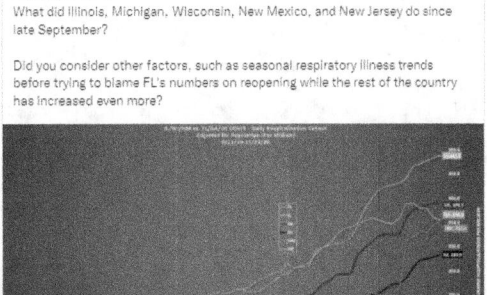

9:57 AM · Nov 23, 2020 · Twitter Web App

TEAM REALITY

Hold2 @Hold2LLC

COVID IL/mi vs. FL/GA: 11/20/20

Now what?

Who's in control? Government or Nature?

Would you rather make your own risk-based choices or have the gov't decide for you?

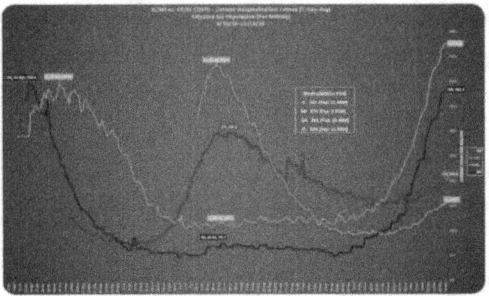

7:00 PM · Nov 20, 2020 · Twitter Web App

Gummi Bear @gummibear737

We should be able to have an honest, open discussion about vaccine injury, without implying that the vaccines don't help people

It's a simple risk/reward analysis

It helps a lot of people based on age, weight and health

It doesn't make sense for my 12/13 year old daughters

1:53 AM · Nov 14, 2021 · Twitter for iPad

Eric @The_OtherET

In Illinois, where they follow the science, they have banned indoor dining unless you bring the indoors outdoors

2:09 PM · Nov 15, 2020 · Twitter for iPhone

Ann Bauer @annbauerwriter

I keep seeing these weepy posts from lockdown supporters about how their favorite cafe/coffee shop/theater is closing. Are they surprised? Is this magic time? Are businesses supposed to enter some fantasy realm where they can just suspend and survive while everyone hides?

12:08 PM · Nov 5, 2020 · Twitter Web App

Martin Kulldorff @MartinKulldorff

"The Zoom call is the 21st century equivalent of the manor estate on the hill, a way to interact with others while avoiding the virus to which the people who keep the goods and services flowing must necessarily be exposed." - @jeffreyatucker aier.org/article/lockdo…

8:02 AM · Nov 2, 2020 · Twitter Web App

Virál Myãlgía MD, PhD @contrarian4data

On Monday, Trump needs to issue an executive order mandating PCR cycle thresholds of 30.

Cases will plummet because primarily true cases will remain.

Whatever happens post-litigation, Fauci-Biden will be forced to accept or reverse and see cases rise.

2:17 PM · Nov 7, 2020 · Twitter for iPhone

Emma Woodhouse @EWoodhouse7

Outdoor restaurant tent at Huron & Wells downtown (Chicago).

Reminder that dining inside the tent is "safe," but dining inside the restaurant building is "unsafe."

Just don't trip over the scaffolding & construction zone barriers as you make your way to your table.

5:46 PM · Nov 18, 2020 · Twitter for iPhone

Gummi Bear
@gummibear737

Public Health may never recover from its failures during this pandemic

Why? - They are making the Exact same mistakes they made during the 1980's AIDS crisis

"Blame and Shame" is NOT how you educate people

1

8:35 AM · Nov 18, 2020 · Twitter Web App

Heavy-handed mandates with no rhyme or reason. Bringing a sledgehammer when a small scalpel is needed. Arbitrary triggers for various NPI's. Why are we still using % positive as a guide for NPIs? (Following the viral levels in sewage waste actually makes more sense.) ++ School closure seems to be a PECULIARLY American thing at this point ++ "Do as I Say...Not as I Do" seems to be rampant among certain mayors and governors. This is not leadership. It's tyranny. ++ Rank Hypocrisy. Certain events are considered "morally" okay....riots and protests for example....while other events are "selfish and stupid"... ++ Abandonment of long-held views on pandemic management. Draconian Lockdowns were frowned upon for good reason. But...We threw away the playbook....and I can't explain why. (This is how conspiracy theories start FWIW) ++ Lack of humility. We even have one governor who has written a book on Leadership despite some really questionable decisions....and results ++ How do we fix this? I don't know. Leaders...state and local...can start by actually communicating with the people they represent. Communication. Truth, Trust, and Transparency. Respect and Humility. Empathy. Leadership by example. Look to the community, not Twitter. My thoughts.

Justin Hart
@justin_hart

The chances of YOU dying from ME not wearing my seatbelt are about the same chances of YOU dying from ME not wearing my mask. Unless you're in the car with me my body is not going to launch in the air and kill you. If I pass you at the store, you ain't gonna die from my #COVID19.

9:45 AM · Nov 27, 2020 · Twitter for iPhone

Bethany S. Mandel ✓
@bethanyshondark

I'm really glad the New York Times etc have joined the reopen schools party. Nothing in the data changed, unless you're counting who got more electoral votes. Nevertheless, welcome. It's too late for this school year, hope it was worth it.

8:13 AM · Nov 19, 2020 · Twitter for iPhone

Jennifer Cabrera 🦔
@jhaskinscabrera

More excellent analysis by @emilyvburns: 'The trade off was never supposed to be, "and for every two 78-year-olds we save, one person with decades left to live will die from despair, or from lack of medical treatment or diagnosis."'

thepragmatist.co/post/more-than…

4:37 PM · Nov 17, 2020 · Twitter Web App

Emily Burns 😷 DMs welcome #TeamReality
@Emily_Burns_V

Apocalyptic predictions are being used to keep kids out of school & businesses closed. This is not the apocalypse. In article: Population-adjusted Cases, Deaths & Hospitalizations for all 50 states vs MA & NY @jhaskinscabrera @nosmhnmh @AlexBerenson thepragmatist.co/post/populatio…

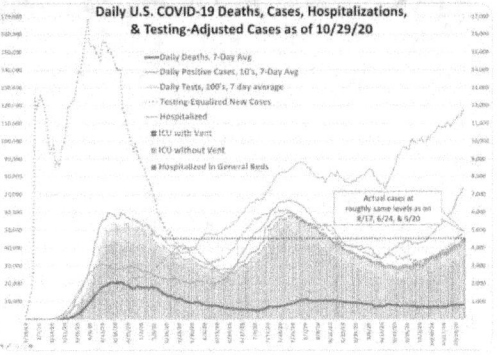

11:31 AM · Nov 1, 2020 · Twitter Web App

Link:
http://www.thepragmatist.co/post/more-than-50-of-u-s-states-never-exceeded-2017-2018-weekly-flu-deaths-state-by-state-analysis

TEAM REALITY

Eric
@The_OtherET

The friends I know that have gotten COVID:

1. Got it while wearing PPE (mask, face shield, gloves)
2. Got it in an office where they are required to wear a mask
3. Nurse, got it with mask/PPE
4. Husband of #3

It's dangerous to tell people they're safe if they just wear a mask.

10:12 AM · Nov 17, 2020 · Twitter for iPhone

Stop blaming people for getting sick. You can do everything right and it still happens. Blaming the public for not "slowing the spread" of a highly contagious respiratory virus accomplishes nothing except make politicians and public health "experts" feel better about themselves. ++ To all the replies saying "no one says this", may I introduce you to Robert Redfield, director of the CDC and the director of my city's local health department, @RexArcherMD who has said you can't infect someone if you have a mask on. At all.

Emma Woodhouse
@EWoodhouse7

Is anyone else experiencing serious Governor envy right now?

Governor Kristi Noem @govkristinoem · Nov 18, 2020
Unfortunately, the spread of #COVID19 is rising in nearly every state, regardless of if they have mask mandates in place. Here in South Dakota, we're focusing on solutions that DO good, not on responses that FEEL good.

1:02 1M views

7:29 PM · Nov 18, 2020 · Twitter for iPhone

Jennifer Sey
@JenniferSey

"... parents with low incomes were 10 times more likely to report that their children were doing little or no remote learning than those making upward of $100,000."

Yes we're losing a generation of kids & I don't get why so few seem to care. nytimes.com/2020/11/06/nyr...

9:11 PM · Nov 6, 2020 · Twitter for iPhone

Eric
@The_OtherET

CDC Director Robert Redfield, September 2020: ""I might even go so far as to say that this face mask is more guaranteed to protect me against COVID than when I take a COVID vaccine"

11:13 AM · Nov 17, 2020 · Twitter for iPhone

Jenin Younes
@Leftylockdowns1

There is no fundamental right not to be exposed to pathogens, much less to force other humans to live in ways that ostensibly reduce your chance of being exposed to them. People understood this throughout human history. Why did this obvious truth go out the window in 2020?

1:16 PM · Nov 28, 2020 · Twitter for iPhone

Zac Bissonnette
@ZacBissonnette

I love Bill de Blasio admitting that he has no evidence tying cases to gyms or restaurants but that if cases keep going up, we have to do something, so he'll close them.

Earnest question: Have we tried sacrificing a goat to the gods of covid and science?

10:14 AM · Nov 27, 2020 · Twitter Web App

TEAM REALITY

DECEMBER 2020

Jennifer Cabrera 😃
@jhaskinscabrera

What does it take to make you angry that politicians who continue to get salaries are closing businesses? A closed business still pays for rent, utilities, insurance - with no income. Their bank accounts DRAIN while politicians break their own rules to avoid any inconvenience.

10:32 AM · Dec 22, 2020 · Twitter Web App

1\ This thread commemorates those brave politicians who didn't let lockdowns slow them down

Send me other examples by DM, and I'll add them

First, Deborah Birx: retiring in disgrace after gathering 2 households and 4 generations for a Thanksgiving party

2\ Then there was Nancy Pelosi, who helped shutter a nation of small businesses before arranging to see her stylist. Hey, at least she liberally disinfected herself with vodka ++ 3\ And who can forget Gavin Nuisance, gleefully presiding over the nation's longest and most severe lockdown, while breaking the law to dine at French Laundy ++ 4\ In LA, Warren Buffett's son dressed in drag, voted to ban outdoor dining, then tottered off to...eat at an outdoor restaurant. What is it with California politicians? ++ 5\ Fauci, ah Fauci. Has any hall monitor ever had sweeter revenge on an entire mocking nation? ++ 6\ Chicago mayor Lori Lightfoot pulling a Pelosi to go get her

hair did. Someone should throw holy water on this one and see if it crawls on the ceiling ++ 7\ Bill DeBacle breaks "stay at home order", walks in park ++ 8\ Chief Mask Nancy, Andrew Cuomo, caught walking dog without mask ++ 9\ Pennsylvania Health Secretary Rachel Levine orders care homes to accept COVID-19 patients, then moves her mother out of a care home ++ 10\ "Professor Lockdown" writes the shitty

model that scares the world into lockdown, then breaks lockdown to shag married woman ++ 11\ Illinois governor Pritzker flies his family out of lockdown to their ranch on private jet: ++ [Thread continues with more examples of politicians breaking their own rules:

https://twitter.com/ElonBachman/status/1341465108736827394]

TEAM REALITY

Michael P Senger
@MichaelPSenger

In the 1300s, as the Black Plague engulfed Europe, cities resorted to a novel public health strategy: killing all cats, which were believed to be conduits for disease and witchcraft.

thevintagenews.com/2017/02/06/in-...

2:16 AM · Dec 18, 2020 · Twitter for iPhone

Tracy Hoeg, MD, PhD
@TracyBethHoeg

Join us physicians in the fight to open schools in the US. We know how bad COVID is, but failing to open public elementary schools - for now nearly an entire year - is creating an even more serious public health crisis that, if you have been watching Europe, is entirely avoidable

> **Vinay Prasad, MD MPH** @VPrasadMDMPH · Dec 7, 2020
> The virus affected the US and Europe, but only the US abandoned kids
>
> The single greatest error in the US pandemic response, and the longest lasting one.
>
> Will scar the nation for decades-- the repercussions so devastating, volatile and unpredictable
>
> washingtonpost.com/education/stud...

7:38 PM · Dec 7, 2020 · Twitter Web App

Jennifer Cabrera
@jhaskinscabrera

COVID is endemic (everywhere) and highly contagious.

It's not your fault if you get it.

It's not your fault if someone gets it from you.

We need to shift from blaming victims to talking about how to devote our resources to protecting the vulnerable.

11:38 AM · Dec 26, 2020 · Twitter Web App

When the culling of cats had the unintended consequence of increasing rodent infestations, worsening the plague, the townsfolk reconsidered and determined their policy was not strict enough, so began killing cat-owners as well. ++ This same civilization—which ultimately lost nearly half its population to plague—also gave us the concept of quarantine, which modern propagandists periodically (erroneously) cite as precedent for our population-wide lockdowns.

Victoria Fox
@drvictoriafox

Wow! NPR tonight. Teachers justifying why it's better for schools to be closed. "Kids are resilient. Parents tell me their kids are falling behind but they have to understand they are falling behind in arbitrary developmental goals we've set". 💀

9:57 PM · Dec 22, 2020 · Twitter for iPhone

AJ Kay
@AJKayWriter

I hereby free you of your obligation to protect me with your mask.

I don't consent to your guardianship of my health.

5:16 PM · Dec 12, 2020 · Twitter for iPhone

Josh Stevenson
@irinadastick

Welcome to 2020, where in order to avert the crisis of healthcare capacity shortages we...

Created capacity shortages.

It was never about the beds.

bls.gov/iag/tgs/iag622...

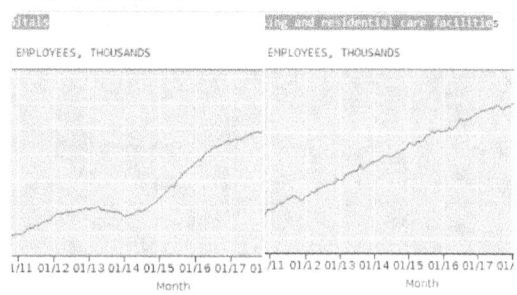

12:17 AM · Dec 7, 2020 · Twitter Web App

TEAM REALITY

Mark Changizi @MarkChangizi

When people believe the sky is falling, don't bother getting advice from the "sky-fall experts" (e.g., weather scientists, astronomers, roofers, etc.).

12:29 PM · Dec 21, 2020 · Twitter Web App

People are always our greatest danger. ++ The scary thing about finding out that the sky is falling is not that the sky is falling, but that everyone believes that the sky is falling. ++ One of the telltale signs of a mass delusion is that one peculiar expertise is suddenly raised to holy status. ++ Civil liberties are only as strong as your neighbors are composed under pressure. ++ That one of the most boring academic fields suddenly became super interesting was itself a sign of mass psychosis. ++ Maybe panic isn't the best medicine. ++ 2020 was the first time an epidemiologist had ever been called to the scene for a case of the cooties. Nearly every epidemiologist showed up. ++ Civil liberties unless "Boo!" ++ It's safer to catch Covid than to get caught dissing Covid. ++ Vitamin D won't protect you from the helpful consequences of mass delusions. ++ Helplessness is when tin foil hats are not only mandatory, but come in a million colors and designs. ++ I prefer liars over the brainwashed. A liar won't destroy you and feel good about it. ++ Your screams only confirm that my cure needs to be administered longer and harder. ++ Never believe that because the mass delusion is so profoundly damaging at all levels, that it will surely stop. A train on fire can go on and on.

Vinay Prasad, MD MPH @VPrasadMDMPH

When Sars cov 2 is over, we will have to deal with the damage done to school kids-- disproportionately poor and minority kids deprived of school for months-- this damage will haunt us for the rest of my life and deeply scar the social fabric.

9:21 PM · Dec 3, 2020 · Twitter for Android

The activists who, at every opportunity, argued for school closure will fail to deliver the remedies needed to correct the damage and their empty words will generate a massive political backlash ++ The damage will be so catastrophic that the fate of our nation may be affected. Democracy and civics will be affected. Feel free to screenshot this tweet. And f/u in 20 years. I wish it doesn't happen but the die is cast. The greatest error of my life has already happened.

John Ziegler @Zigmanfreud

Even as pessimistic as I am, I REALLY doubted that King @GavinNewsom would have the audacity to lockdown CA again, especially after being caught in scandal & during Christmas, but it is becoming increasingly obvious he will do it because he knows the media/public will not revolt.

12:01 PM · Dec 2, 2020 · Twitter for iPhone

Clifton Duncan: Good Looking Loser. @cliftonaduncan

The fear-mongering won't stop at Covid; the precedent's been set:

We've shown that not only will we blindly obey, but we'll demonize & rat out neighbors, abandon what makes life vital, and ruin our children's futures at the drop of a hat, as long as we're told it's "science".

1:37 PM · Dec 22, 2020 · Buffer

ClownBasket @ClownBasket

And now...the 2020 Reality Awards!

Celebrating the men, women and bots of Team Reality who have so masterfully covered the SARSCoV-2 virus, Covid-19 disease, and related hysteria, corruption and heroics.

A tribute from the "little" accounts who found community during chaos.

4:50 PM · Dec 31, 2020 · Twitter Web App

Stinson Norwood @snorman1776

There is *some* justification (as small as it may be) for healthcare workers getting the vaccine before Grandma.

There is NO justification for a previously infected healthcare worker getting the vaccine before Grandma. A more selfish person does not exist.

1:32 PM · Dec 24, 2020 · Twitter for iPhone

AJ Kay
@AJKayWriter

Dr. Birx's fatal flaw was not her need to interact with her family.

Human-to-human, I'm glad she refused to abandon those connections.

My outrage is squarely focused on any policy/position/politician who regards physical interactions w/ loved ones as frivolous or unnecessary...

5:46 PM · Dec 22, 2020 · Twitter for iPhone

While her hypocrisy is infuriating, it stands as a crystal clear illustration that policies of extended social isolation are unhealthy. Dr. Birx's struggles were well-articulated in today's NY Post article announcing her retirement— ++ Dr. Birx-type bureaucrats come and go. People in positions of power abuse it. Such is the nature of humanity, TBH. Hypocrisy abounds and I'm not suggesting that we stop calling it out. Make no mistake, nothing happened *to* Birx except that which she called down on herself. ++ The brave position would've sounded something like "I believe that limiting gatherings could help stop the spread—but I understand that many people, like me, are in need of their loved ones support. Here are some things to consider..." But other motivations overrode honesty... ++ Should she resign? Absolutely. But, IMO, not because she snuck around to see her family—even after telling others not to. She should resign because she promoted the policies that required her to sneak around in the first place. Her positions were anti-public health. ++ Our rage should be trained on this idea that every individual should be forced to sacrifice their humanity & stop their lives to hide from an already widespread, overwhelmingly survivable respiratory virus —with zero regard for the ever-growing list of known harms that result. ++ Birx failed b/c—absent human rights violating enforcement—humans will always fail at mandated social isolation in the absence of proportionate threat. The costs are too high. & anyone who doesn't understand that fact, IMO, isn't qualified to advise on matters of public health. ++ A truthful apology wouldn't have said, "I'm sorry I broke the rule" But, instead, "I'm sorry I made the rule."

Stinson Norwood
@snorman1776

Find some cashiers and truck drivers and service staff, and if you can, give them something. They're the true "frontline" workers, the true heroes—-many of whom never got furloughed or had hours cut, so they never got additional money from the state.

10:08 AM · Dec 12, 2020 · Twitter for iPhone

Eli Klein
@TheEliKlein

NY has 25% more current Covid hospitalizations per capita than FL and NY is worse than FL in every other cumulative and current Covid-related metric

FL is almost completely open and normal

NY is very much closed and irrational

Nobody I talk to in NYC knows that data, nobody

7:57 PM · Dec 14, 2020 from Manhattan, NY · Twitter for iPhone

Martin Kulldorff
@MartinKulldorff

The #GreatBarringtonDeclaration now has over 750,000 co-signers from around the world, supporting focused protection instead of #lockdowns. Add your name! gbdeclaration.org

tizens	medical & public health scientists	medi
)9	12,906	

9:57 AM · Dec 17, 2020 · Twitter Web App

TEAM REALITY

Matt, Pre-School Diploma
@statomattic

"Denier" is an ugly slur meant to evoke the Holocaust. But if someone calls you a denier b/c you won't subjugate your life to a virus, embrace it.

Deny that science is a political weapon rather than a method of inquiry.

Deny that media propaganda will determine your actions.

8:43 AM · Dec 9, 2020 · Twitter for Android

Deny the corrosive fallacy that harmful government mandates can defeat a virus. Deny that this virus is so different from all others before that we must throw decades of scientific consensus to the wind. Deny the myopic viewpoint that pretends tradeoffs don't exist or matter. ++ Deny the lie that denying children who are not at risk precious formative months and years of education is EVER okay. Deny the dehumanization of relegating our vital social events to the flattening simulated world of Zoom. Deny that the Uber Eats driver's health matters less. ++ Deny that there is zero risk in performing any activity, any day, in any era. Deny that holidays and special occasions with our families can simply be canceled because they are certain to happen next year anyway. Deny that these gatherings are trivial and selfish in any way. ++ Deny that a virus with a 99.8% survival rate should, in any remotely sane world, necessitate cooping healthy, low-risk people up in their homes. Deny that this isn't inhumane. Deny, deny, deny, deny. If that's what being a denier means today, sign me up.

John Ziegler
@Zigmanfreud

I would pay good $ for a rational/fact-based explanation for why, when you combine the COVID data from California, Illinois & New York, it is, at best, no better than doing the same for Florida, Texas & Georgia.

If lockdowns work even a little bit, this should NOT be possible!

7:28 PM · Dec 20, 2020 · Twitter for iPhone

Prof Francois Balloux
@BallouxFrancois

I'm a bit puzzled that some still deny the concept of 'herd immunity' despite the incoming vaccines. Please let me state this one more time, there's nothing political about herd immunity, it's not a strategy nor a policy, it's just a fact of life.
1/

> **Prof Francois Balloux** @BallouxFrancois · Sep 30, 2020
> Herd immunity (community immunity) is not a strategy. It is the process describing that not everyone in the population needs to be immunised before an epidemic recedes. It is at the heart of any vaccination campaign, as vaccines are generally not indicated for everyone.
> 1/ twitter.com/devisridhar/st...
> Show this thread

4:50 PM · Dec 18, 2020 · Twitter Web App

Jenin Younes
@Leftylockdowns1

I disagree with those who say, just wear masks because why not? Setting aside the harm, it's dangerous to allow gov't to force us to do things without evidence they effectuate the goal. That gives it carte blanche to make us do anything to give the appearance of doing something.

9:37 AM · Dec 24, 2020 · Twitter Web App

Justin Hart
@justin_hart

The mind-twisting logic of the FauciBirx brigade: "Trust FauciBirx! FauciBirx IS science!"
Also: "Everything President Trump did was a mistake when it comes to the pandemic."

Um... Trump did everything Fauci & Birx asked... Get some help folks.

9:50 AM · Dec 22, 2020 · Twitter Web App

Stinson Norwood
@snorman1776

January: "Trump is a fascist. We have to get rid of him by any means necessary."

May: "I wear my mask for you."

November: "We're finally going to have a President who cares about us and follows science."

December: "Papers, please."

10:13 AM · Dec 6, 2020 · Twitter for iPhone

TEAM REALITY

Martin Kulldorff
@MartinKulldorff

Twelve Forgotten Principles of Public Health

#1 Public health is about all health outcomes, not just a single disease like #COVID19. It is important to also consider harms from public health measures.
#totalharms
collateralglobal.org @collateralglbl

12:45 PM · Dec 19, 2020 · Twitter Web App

#2 Public health is about the long term rather than the short term. Spring #COVID19 #lockdowns simply delayed and postponed the pandemic to the fall. ++ #3 Public health is about everyone. It should not be used to shift the burden of disease from the affluent to the less affluent, as the #COVID19 #lockdowns have done. ++ #4 Pubic health is global. Public health scientists need to consider the global impact of their recommendations. ++ #5 Risks and harms cannot be completely eliminated, but they can be reduced. Elimination and zero-COVID strategies backfire, making things worse. ++ #6 Public health should focus on high-risk populations. For #COVID19, many standard public health measures were never used to protect high-risk older people, leading to unnecessary deaths. ++ #7 While contact tracing and isolation is critically important for some infectious diseases, it is futile and counterproductive for common infections such as influenza and #COVID19. ++ #8 A case is only a case if a person is sick. Mass testing asymptomatic individuals is harmful to public health. ++ #9 Public health is about trust. To gain the trust of the public, public health officials and the media must be honest and trust the public. Shaming and fear should never be used in a pandemic. ++ #10 Public health scientists and officials must be honest with what is not known. For example, epidemic models should be run with the whole range of plausible input parameters. ++ #11 In public health, open civilized debate is profoundly critical. Censoring, silencing and smearing leads to fear of speaking, herd thinking and distrust. ++ #12 It is important for public health scientists and officials to listen to the public, who are living the public health consequences. This pandemic has proved that many non-epidemiologists understand public health better than some epidemiologists. / END

Phil Kerpen
@kerpen

New JAMA meta-analysis of 54 studies with 77,758 participants finds household secondary attack rate (chance an infected person will infect one or more people at home) is 18% if the index case is symptomatic and 0.7% if asymptomatic.
jamanetwork.com/journals/jaman...

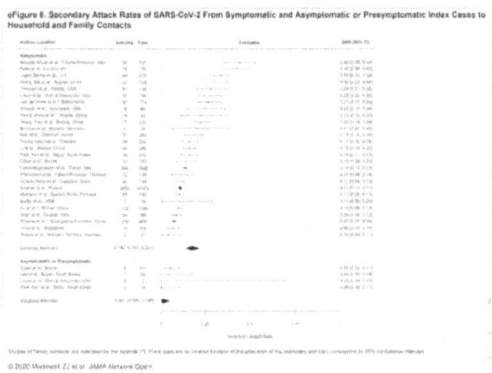

3:33 PM · Dec 14, 2020 · Twitter Web App

Woke Zombie
@AWokeZombie

Cuomo shows the following chart. He conflates Social gatherings with Household spread instead of splitting them out (Wonder why? - in April it was 80% household/0 SG)

10 mins later: Deblasio closes indoor dining.

Normal people: "Wtf?"

When will people rise up?

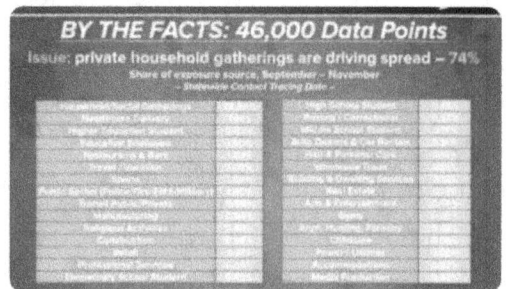

12:13 PM · Dec 11, 2020 · Twitter Web App

TEAM REALITY

Prof
@covidtweets

Reflections on 2020...

Whoever you ask, regardless of their views on lockdowns, masks, etc., they will say our response to COVID was a disaster.

How did we manage to get this much wrong?

My few cents: (1/x)

9:47 PM · Dec 30, 2020 · Twitter Web App

Reason #1: Social media/media. After (likely staged) scenes from China, misinformation resulted in the fear quickly taking over. When fear is prevalent, most people cannot reason. Quickly, people began turning to knee-jerk reactions. ++ Reason #2: The risk of severe illness is not equally distributed. Same for lockdowns, which also affect people differently. The problem is, there is not a big overlap between those with the short end of the stick from COVID vs. lockdowns. ++ Reason #3: Social media/media. This unequal stratification of risk quickly resulted in two camps emerging. Once this happened, our innate desire to tribalize made things go downhill, resulting in less and less debate and more and more blaming/name-calling/attacks. ++ Reason #4: The safety/security that came with modernity made people forget what it was like to move forward with life despite the risks. It had been decades since the last time there was such a major scare. People lost perspective. ++ he thought of dying of something one cannot stop/control goes against our innate survival instincts. However, it is a normal part of living. We just don't think about it much. Most people currently alive forgot what it was like to live this way. ++ Reason #5: Social media/media. Constant coverage/discussion of COVID made sure people never stopped fearing the possibility of death. Even though the risk was much lower for most people, once fear took over, facts did not matter. ++ Reason #6: Election. The fact that all this happened in an election year, and the US political scene being divided on the importance placed on personal liberties, amplified all of the above. It became impossible to get people agree with each other on anything. ++ Reason #7: Social media. Even in non-pandemic election years, social media is incredibly toxic and polarizing. With COVID, it became exponentially more so. Everything that went wrong became Trump's fault. His overall attitude and language did not help either. ++ Reason #8: Human psyche. Once someone picked a side and began acting accordingly, and accumulated the costs, it became increasingly unlikely that new data would change behavior. Governors left and right fell to this trap, and then doubled down. ++ Reason #9: Social media/media. Those craving dopamine amplified even rare events to get interactions. Media needing clicks did the same. Sensationalism ran rampant. ++ ... ... Reason #317: Social media/media. I believe social media, despite its positives, is the biggest threat to our existence. I don't know the solution. ++ Could things be different? Maybe if China did not pump propaganda in February/March, and Italy was able to stay calm... Otherwise I don't see how we could have avoided all this. Once it began, it was the perfect storm, with stars aligned exactly as they needed to. ++ I am looking forward to calm discussions among scientists from various disciplines on how we can do better next time. And make no mistake, there will be a next time.

TEAM REALITY

Ann Bauer
@annbauerwriter

It's become clear to me that about 50% of the people I know actually like the sense of being controlled/locked down and supposedly 'safe.' This is astonishing.

I keep waiting for the uprising, the I've-had-enough. But it's not coming. Just the opposite.

2:48 PM · Dec 15, 2020 · Twitter Web App

Friends who initially rankled at the prohibitions now give me a dead-eyed stare (over zoom) when I mention going out in public, sending kids to school, moving freely. They tell me that sounds 'dangerous.' This is a known psychological syndrome, right? Someone, fix it, please. ++ It's really scaring the hell out of me. These are smart people who used to have lives. One-time world travelers. Some have tons of money. They appear brainwashed now. They like the confines of their homes. ++ I'm seriously looking for other people who don't want that safe, controlled life and cannot find it in my city/state/peer group. Where is it? Is there a club? Are they in other states? Idaho...Texas? Maybe some Pacific island? ++ I'm at the point where I'd spend all my savings to get myself and my family to wherever that free place is. I wonder if this is how my grandmother felt fleeing Russia at 15, leaving her family, huddling in the hull of a ship for weeks. I get it now, Sadie. I understand.

Gummi Bear
@gummibear737

Masks can paradoxically be counterproductive with Covid-19

Once you gain confidence in the efficacy of the mask, you will be less worried about socially distancing

I see many mask wearers behaving as if their mask confers immunity

Masks have limited utility at best

4:09 PM · Dec 17, 2020 · Twitter for iPad

Megan "Hey it's the mask lady" Mansell
@mamasaurusMeg

Here's the bottom line:

If you educate the public that they are safe using cloth and surgical masks with airborne contagion, they will continue using the wrong PPE in event of a truly deadly airborne contagion.

That's why we must fight misinformation.

Lives depend on truth.

4:48 PM · Dec 14, 2020 · Twitter for iPhone

Emma Woodhouse
@EWoodhouse7

Vaccination distribution across age groups in Chicago is...puzzling. COVID deaths of elderly sick folks have been used to justify business & school closures, while the same population is evidently not a vaccination priority.

Way to go @chicagosmayor

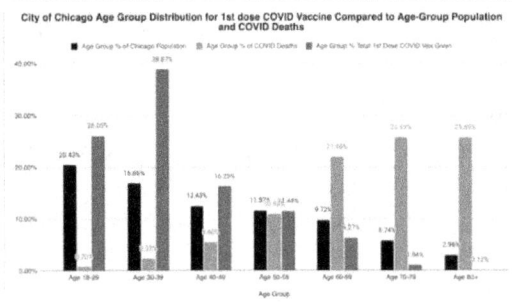

2:39 PM · Dec 26, 2020 · Twitter Web App

Eric
@The_OtherET

3 weeks removed from when people started gathering for Thanksgiving, let's see how those #COVID19 "surge upon a surge" predictions by Fauci etc panned out for those of us in the Midwest. Colored lines indicate each state's peak. They all occurred within one week of each other.

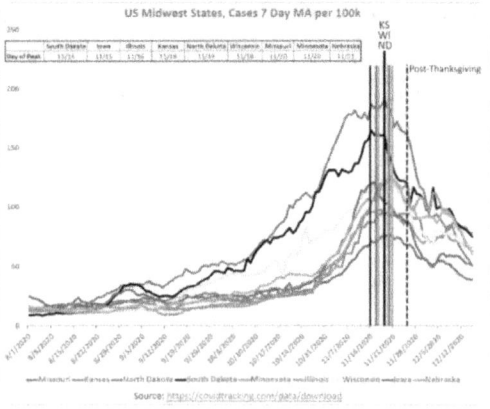

5:35 PM · Dec 16, 2020 · Twitter Web App

TEAM REALITY

Bethany S. Mandel @bethanyshondark

Why aren't people angrier about this

> My 85 year old mother has been stuck at home since March, and she's deteriorated both physically and cognitively. But yes, please give shots to orthodontic receptionists. And to addicts per Comrade Wilhelm.

1:40 PM · Dec 29, 2020 · Twitter for iPhone

Gummi Bear @gummibear737

Joe wants to take credit for the seasonality of COVID-19

Coronavirus infections peak around January, so Joe can tell people to wear blue socks and cases will still decrease

Joe also taking credit for vaccines developed under Trump

School reopenings? So it was safe all along...

> **Joe Biden** @JoeBiden · Dec 8, 2020
> United States government official
> Today, I'm announcing key COVID-19 priorities for the first 100 days of my administration:
>
> - Everyone wears a mask
> - 100 million vaccinations
> - Reopen the majority of schools
>
> With these steps, we can change the course of the disease and change life in America for the better.

10:23 AM · Dec 9, 2020 · Twitter for iPad

Emma Woodhouse @EWoodhouse7

2020 Person of the Year isn't Tony "Goal Post Changer" Fauci, Deborah "My Policies Were Starving My Own Parents" Birx, or Attention-Seeking Staff in Normal-Capacity Hospitals.

It's the Low-Wage "Essential" Worker Single Mom with Kids at Home.

Where's her standing ovation?

1:44 PM · Dec 26, 2020 · Twitter for iPhone

Gummi Bear @gummibear737

Ignoring repurposed drugs/supplements is one of the most egregious healthcare travesties ever

Maybe they don't work (HCQ, Ivermectin, Zinc, Bromhexin, Vit C/D, Zinc), but would have been easy to test

They made sure it was vaccine or bust

2:10 AM · Dec 21, 2020 · Twitter for iPad

Mark Changizi @MarkChangizi

That hysteria, panic and the madness of crowds is our most dangerous enemy needs to be taught.

It is apparently not obvious.

> **Mark Changizi** @MarkChangizi · Dec 16, 2020
> [1]
> "In reading the history of nations, we find that, like individuals, they have whims & peculiarities; their seasons of excitement & recklessness when they care not what they do. We find that whole communities suddenly fix their minds upon one object, & go mad in its pursuit"
> Show this thread

12:17 PM · Dec 28, 2020 · Twitter Web App

Erich Hartmann @erichhartmann

Small business owners, restaurants and performing artists:
WTF are you doing?
Government will not save you.
You have to act.
You have to stick up for yourself.
Now.

10:33 AM · Dec 29, 2020 · Twitter Web App

Eric @The_OtherET

How much longer will people be willing to surrender their rights not even for their safety, but for the illusion of it?

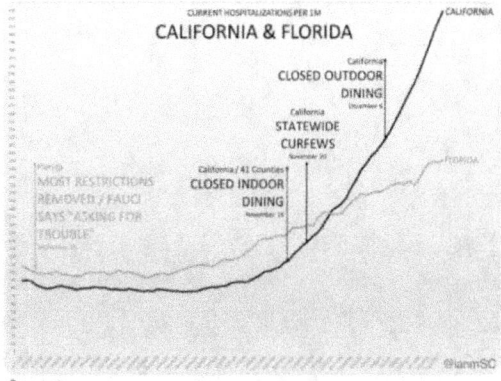

2:24 PM · Dec 19, 2020 · Twitter for iPhone

JANUARY 2021

Martin Kulldorff
@MartinKulldorff

Back in March, there were no evidence that lockdowns would work to protect older high-risk people. Now there is evidence. They didn't.

6:10 PM · Jan 21, 2021 · Twitter Web App

TEAM REALITY

Federico Andres Lois @federicolois

Controversial opinion: if you really care about the elderly and at risk, you wouldn't permit to give the vaccines to not at risk groups in order to diminish the immunological pressure that will guarantee vaccine evasion mutants. That's it I said it.

4:20 PM · Jan 21, 2021 · Twitter for Android

BOUTROS 木 @boutros555 · Jan 29, 2021
Replying to @dsmithmph and @Jim_Jordan

I support your right to wear two masks.

I support my right to wear zero.

Recommending two is a tacit admission that one isn't as effective as our experts told us. Which means we never had the right to force it on to people.

Agreed?

Don Wolt @tlowdon

NYTimes paints another frightening picture of SoCal hospitals in a COVID-driven "crisis," but LA DHS data tells a different story. LA Cnty overall hospital occupancy is down 3331 beds (22%) from its July peak & is mostly flat since Thanksgiving - not what we'd expect in a crisis.

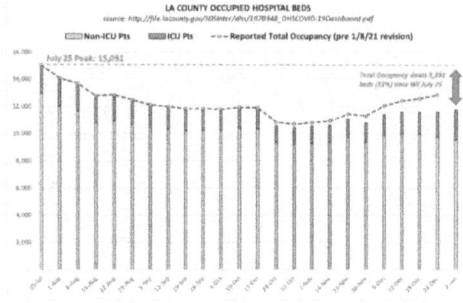

The New York Times @nytimes · Jan 9, 2021

Real Developments @pdubdev

Twitter suspended @boriquagato for using facts.

Dangerous times...

6:02 PM · Jan 5, 2021 · Twitter Web App

Woke Zombie @AWokeZombie

#NJ #COVID19 When will the press ask the following:

"What metrics are required to return to normal? To end 25% dining? To end the public health emerg?"

"What is required to stop wearing masks in schools and in stores in NJ?"

"What's the metrics for unwinding hybrid in schools?"

6:24 PM · Jan 25, 2021 · Twitter Web App

Erich Hartmann @erichhartmann

This 🎥 originally aired in early July 2020. We knew way back then that kids and schools should be 100% open. Every pediatrician in this clip says YES! to #openschools and yet we kept kids out of school this entire time, causing massive, unnecessary long term damage. Shame on us.

10:52 PM · Jan 23, 2021 · Twitter Web App

Ian Miller @ianmSC

Cases in Japan have still not come back down, 3 months after the big increase started. I honestly had no idea there were so many people there who are so selfish that they won't just wear a damn mask

Especially after Vanity Fair & NY Times said months ago that masks worked there

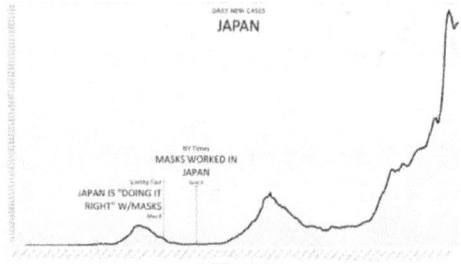

1:32 PM · Jan 21, 2021 · Twitter for Mac

Victoria Fox @drvictoriafox

My family is now 1:1: A COVID related death & a lockdown related death. I don't blame anyone for either but can we please stop denying/ignoring the costs of lockdowns. COVID is real & awful & requires a response. Lockdowns are not the correct response.

10:48 AM · Jan 15, 2021 · Twitter for iPhone

Emma Woodhouse @EWoodhouse7

Illinois has been under a mask mandate since May. My city (Evanston) since April 20.

They don't work.

100 more days won't change that.

2:17 PM · Jan 20, 2021 · Twitter for iPhone

TEAM REALITY

Brumby
@the_brumby

Lockdowns don't actually work (see pinned tweet). One reason is that it's preposterous to think one can just halt an interwoven societal fabric. So "lockdown" is instead just poor people risking exposure to continue providing goods/services to rich people sheltering in place.
1/

1:39 PM · Jan 21, 2021 · Twitter Web App

A thousand books can (and will) be written on the harms of lockdowns. This thread will instead focus on its epic exacerbation of the divide between rich & poor. Last week I let 30 studies do the talking; this time I will let 23 pictures tell the story. ++ 1. High earners are the ones that can work from home. ++ 2. Similarly, only a majority of those with a college or postgraduate degree can work from home (also note the disparity in race/income). ++ 3. This is reflected in mobility by income bracket – those with the higher incomes are able to "stay safe, stay at home". ++ 4. Low wage workers are lucky to have

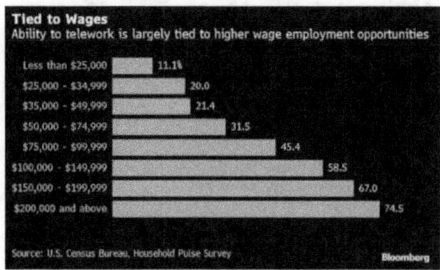

a job at all--remote or otherwise (observe that high earners are above prior peak). ++ 5. Note, even though high earners are now fully employed, total unemployment is still well above the prior PEAK of the Great Financial Crisis. Remember how horrible that was? Low earners are orders of magnitude worse off than during the GFC. Its literally off the prior charts. ++ 6. But didn't we pass a stimulus bill with $600 checks? Yes, that will last less than a month for a majority of recipients. ++ 7. All this increased exposure of poor/racially disadvantaged is borne out in data on hospitalizations/mortality. We all know Los Angeles is a COVID hotspot, but did you know 71% of cumulative LA County Department of Public Health hospitalizations have been amongst Hispanics? ++ 8. This is supported by Latinos having much higher risk of death overall in California, across all ages, and especially acute amongst immigrants. Note also the correlation amongst those lacking education and essential workers. ++ 9. In Toronto, COVID19 mortality was multiples higher amongst the 30 lowest income neighborhoods vs. the 30 highest. ++ 10. Downtown offices are empty, meaning that urban ecosystems catering to office workers have been hollowed out (think taxis, gyms, coffee shops, restaurants & bars, dry cleaners, etc.) crushing minority-owned small businesses. Most downtowns are nothing but boarded up shells. ++ 11. Suburban retail traffic has likewise dropped, but to a staggeringly smaller degree. ++ 12. Minority-owned small businesses have been crushed to a much greater extent than white-owned small businesses. ++ 13. This is also reflected in delinquent mortgage payments being more prevalent in predominantly Black counties (e.g., Clayton in Georgia is >70% Black and has triple the rate of Forsyth (mostly white)) ++ 14. The &P500 index (large companies) is at an all-time high; small business revenue is still down 30%+ ++ 15. For the first time in decades, the World Bank is predicting an increase in extreme poverty. Recall also the UN predicts lockdown will push 130 million to the brink of starvation. ++ 16. Female labor force participation is dropping, wiping out decades of gains. Typically below the rate for males, female unemployment is now notably higher. ++ 17. Sadly, this is likely due in large part to school closures. In Calif, its almost exclusively private schools that are open. We live in a mostly meritocratic society, but where there are many structurally disadvantaged groups, in large part because they haven't been given the tools (education) to succeed. How will two-years of no-school impact that California? Who do you care about? ++ 18. So that is life for the poor. I know its coming, so for those that would argue the virus closed the economy, not the lockdowns, Gov. Kristi Noem has a message for you: ++ 19. How about life for the well-to-do, you ask? Well, net worth for those that own assets is at an all-time high, eclipsing pre-pandemic levels ++ 20. Home prices have skyrocketed post-lockdown and are at an all-time high as people flee tyrannical urban areas (further hurting the inner-city poor) ++ 21. Mortgage rates have collapsed, meaning wealthy homeowners can extract this increased value from their homes at no cost as monthly debt service payments cratered. ++ 22. FANMAG stocks (Facebook, Apple, Netflix, Microsoft, Amazon and Google) have exploded post lockdown and are at all-time highs and cash flow metrics. ++ So why do "progressive" Dems support these policies? If you support them, where do you fall in the spectrum above? Biden's team will "establish a national covid-19 response structure where decision-making is driven by science and equity" —I'll save you the effort #endthelockdowns.

Governor Kristi Noem @govkristinoem

GREAT NEWS: South Dakota's unemployment rate is down to 3.5% for November. In November 2019, it was 3.4%.

7:12 AM · 12/17/20 · TweetDeck

Tracy Hoeg, MD, PhD @TracyBethHoeg

From our research group. Published today by @CDCgov . Rural WI, 5530 K-12 students and staff over 13 weeks; comm pos rate up to 40%. Only 7 cases of in school transmission, all in children. There is no apparent reason schools should be closed.

cdc.gov/mmwr/volumes/7... via @CDCgov

cdc.gov
COVID-19 Cases and Transmission in 17 K-12 Schools ...
The coronavirus disease 2019 (COVID-19) pandemic has disrupted in-person learning in the United States, with approximately one half of all students ...

2:23 PM · Jan 26, 2021 · Twitter Web App

Kyle Lamb @kylamb8

In general, there has never been an argument on the right side of history that required censorship to achieve the position. If your position requires suppressing discussion or opposition, it's not very strong.

1:24 PM · Jan 13, 2021 · Twitter Web App

And, like magic (just as I predicted) the news media at the White House is dutifully taking their marching orders from the @nytimes & now 2X virtue-signaling by DOUBLE masking!

Glad they're admitting a mask doesn't work & that the "experts" have been totally wrong for 11 months!

2:09 PM · Jan 25, 2021 · Twitter for iPhone

TEAM REALITY

John Ziegler @Zigmanfreud

Zoom "schooling," at least for our 2nd grader, is now actually worse than 0 school at all.

Not only is there 0 learning, & she's being taught that school clearly is not that important, but now she's constantly in tears of frustration & HATES "school."

Nice job teacher's union!

12:28 PM · Jan 8, 2021 · Twitter for iPhone

Mark Changizi @MarkChangizi

The lockdowners believe that the debate is about the science. It's not. The real debate concerns the lockdowners' mistaken belief that,

If they're right about the science, then it justifies violating human rights.

4:45 PM · Jan 6, 2021 · Twitter Web App

Vanessa @vla42

Presented without commentary.

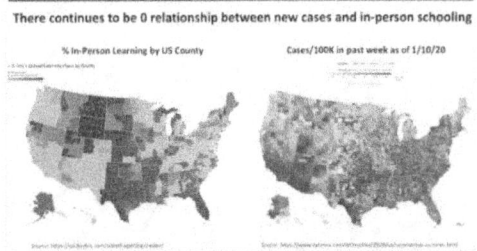

2:01 PM · Jan 14, 2021 · Twitter Web App

Emma Woodhouse @EWoodhouse7

Peak virus paranoia irony in my city today:

People sledding on a forbidden hill that has danger signs all over it…with face masks on

Acceptable risk: Breaking your neck
Unacceptable risk: Being outside maskless

3:15 PM · Jan 27, 2021 · Twitter Web App

Wes Pegden @WesPegden

When the history of extended school closure is written, it will not say that "even in Jan 2021, no one knew if school could operate without frequent outbreaks".

It will be said that we did know, and looked the other way; that 2 worlds existed: one where children were cast aside.

11:58 PM · Jan 26, 2021 · Twitter for Android

Justin Hart @justin_hart

We've lost our ever-lovin' minds.

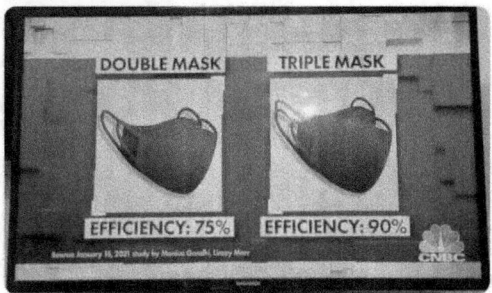

7:13 PM · Jan 26, 2021 · Twitter Web App

Vinay Prasad, MD MPH @VPrasadMDMPH

If NYTimes added to their deaths counter w/ a running counter of the kids who are being abused, losing grade-levels, with new onset depression or suicidality, or gunshot injury or death

AKA made the trade off emotional available

Public sentiment would shift overnight

9:18 PM · Jan 18, 2021 · Twitter Web App

Eli Klein @TheEliKlein

In Florida, January Covid vaccination appointments for people 65 and older are simply booked online

In New York, vaccinating people 65 and older "may result in a penalty to the provider of up to $1 million, and revocation of all state licenses"

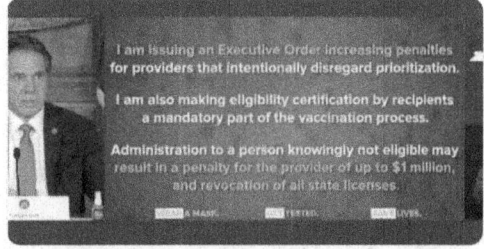

8:09 PM · Jan 3, 2021 from Manhattan, NY · Twitter for iPhone

TEAM REALITY

Megan "Hey it's the mask lady" Mansell
@mamasaurusMeg

Folks say, "Ohhh but a mask will shorten the distance traveled by a particle," overlooking two key factors:
• masks aerosolize droplets that would've fallen in a 6 foot arc into an 18-20 foot range
• these tiny particulates don't respond predictably to gravity

So effectively 1/

9:42 AM · Jan 25, 2021 · Twitter for iPhone

You've taken something with predictable behavior, one form of Covid transmission (droplet) and turned it into the transmission route that is far more difficult to mitigate, and sent it aloft where it will remain for extended periods. Am I the only one who sees a problem here? ++ P.S. The vast majority of masks do not filter the aerosolized Covid particulate range on inhale or exhale, right within respiratory range, while still commingling vulnerable with mixed populations and calling it a day, for a pathogen with minimal viral load for transmission.

AJ Kay
@AJKayWriter

At some point, there needs to be an audit of the 400K US deaths attributed to #COVID19, as well as how the reporting guidelines for C19 impacted that total.

There's far too much at stake & too many variables at play to not attempt an honest accounting of who died of what & why.

7:07 PM · Jan 23, 2021 · Twitter for iPhone

Stinson Norwood
@snorman1776

Imagine in March of 2020 if we told people:

1) Stay home if you don't feel well
2) Stay home if someone in your family doesn't feel well
3) 65+ & provider-endorsed immuno-comp: stay home - here's money for you
4) Check in on 65+ neighbors but keep your distance

That's it.

5:12 PM · Jan 21, 2021 · Twitter Web App

ClownBasket
@ClownBasket

Replying to @FatEmperor

Back in April John Giesecke from Swedish team was taking arrows for their "plan". This interview really opened my eyes to decision making and cost/benefit.

When asked about neighboring countries he said, "let's see in a year."

Wow was he right.

unherd.com
Swedish expert: why lockdowns are the wrong policy
That was one of the more extraordinary interviews we have done here at UnHerd. Professor Johan Giesecke, one of the world's most senior ...

3:09 PM · Jan 25, 2021 · Twitter for iPhone

Martin Kulldorff
@MartinKulldorff

The lockdown strategy is focused protection of affluent professionals, while letting it rip among the working class, hitting inner cities the hardest.

> Martin Kulldorff @MartinKulldorff · Dec 19, 2020
> #3 Public health is about everyone. It should not be used to shift the burden of disease from the affluent to the less affluent, as the #COVID19 #lockdowns have done.
> torontosun.com/opinion/column...
> Show this thread

5:45 PM · Jan 15, 2021 · Twitter Web App

Prof Francois Balloux
@BallouxFrancois

This is my last tweet for a while about lockdowns. Lockdowns are blunt, brutal, discriminitarory, unfair, largely ineffective measures. They can only be seen as an abject failure in planning, but there's a point where they are the only option on the table, as right now in the UK.

1:18 PM · Jan 6, 2021 · Twitter Web App

Prof Francois Balloux
@BallouxFrancois

I'm seeing claims more transmissible / lethal #SARSCoV2 variants emerged because we didn't suppress it hard enough, or conversely, because we suppressed it too hard. Both arguments can be made to work in toy models, but neither is overly convincing with realistic parametrisation.

3:48 PM · Jan 22, 2021 · Twitter Web App

Ann Bauer
@annbauerwriter

Last spring, I looked for high-quality anti-lockdown accounts to follow. The pp I found fall mostly into these categories:

1. Mathematicians & statisticians
2. Futurists, specialists in AI
3. Academics/writers who research mass hysteria
4. Doctors who aren't epi/ID

10:50 AM · Jan 16, 2021 · Twitter Web App

Sure, there's the random chef, stay-at-home mom, artist. (I avoided the MAGA don't-tread-on-me crowd) But I find it interesting that the people who have made their careers studying facts & stats, breakthrough tech and general health are aggressively represented. ++ I happen to be married to someone from category 1. Since March, he's been adding up numbers, comparing stats, recalculating everything he sees on CNN. He cannot help himself. Since March I've been hearing him say the same thing: This makes no sense. The numbers don't add up. ++ Every time I start to doubt myself I log on and see a post from someone who has spent their professional life making rational judgments based on objective data. They all say this is a bubble, a mass delusion. The disease is real but the narrative is not.

Dr Clare Craig (not one of her impersonators)
@ClareCraigPath

The 45,000 extra homeless people will each have a reduced life expectancy of 17.5 years on average.

That is 787,500 years of life lost thanks to that one aspect of collateral damage.

theguardian.com
Tens of thousands made homeless despite UK ban on evictions during pandemic
Charities say younger people working in hospitality among worst affected

3:35 PM · Jan 5, 2021 · Twitter Web App

Eric
@The_OtherET

Kansas contact tracing data shows 0.91% of cases can be traced back to restaurants and bars, with a whopping 0 deaths resulting from them

9:28 PM · Jan 13, 2021 · Twitter Web App

Emma Woodhouse
@EWoodhouse7

"Private companies can do what they want"

Excellent.

Open up, Chicago bars & restaurants.

1:00 PM · Jan 9, 2021 · Twitter for iPhone

Infectious Disease Ethics @ID_ethics · Jan 19, 2021
We also need to be careful when assuming that reducing the transmission of viruses in the short term constitutes a long-term public health benefit

More people susceptible to common viruses in future years may well lead to larger seasonal epidemics & more harm in high risk groups

TEAM REALITY

Bethany S. Mandel ✓
@bethanyshondark

This is the pandemic of privilege. People with money can:
- Use the poor as human shields, doing all of their "essential work"
- Send their kids to private schools, homeschool with tutors, etc.
- Obtain private special services (therapies, etc) privately & pay out of pocket

5:28 PM · Jan 24, 2021 · Twitter Web App

If you are not loudly advocating for the reopening of schools and society in general, you never again get to tell anyone to "check your privilege." ++ I count myself as one of the privileged. I use Instacart (and always have), I am paying $$$ for private occupational therapy because everything public & covered in-network is via Zoom and inadequate. I am spending A LOT of time forming social groups for my kids. ++ I have been loudly in favor of reopening because I recognize that privilege and see how damaging this is for those without the opportunities we have.

Eric
@The_OtherET

Tired of seeing the same flawed #COVID19 Kansas mask study being shared as proof of "masks working", so I used similar methods + the same dataset and tracked what happened in mandate/non-mandate counties after their study's end date. Pretty different from what the CDC found.

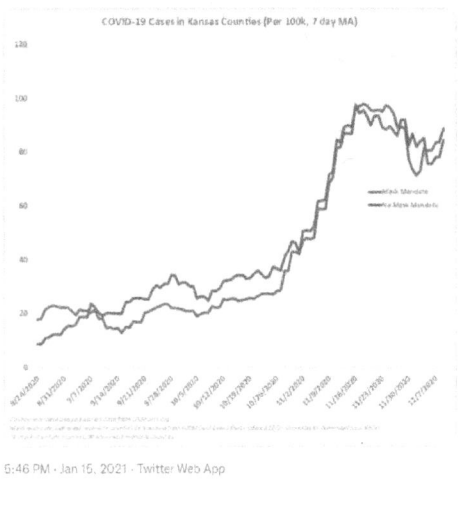

5:46 PM · Jan 15, 2021 · Twitter Web App

Vinay Prasad, MD MPH ✓
@VPrasadMDMPH

The idea there is something called 'zero risk' is a mirage, a delusion, but unfortunately this has gained traction

It prevents us from making the hard and important call on schools, and precludes appropriate post-vaccine messaging

Life has risks, and policy is about tradeoffs

11:57 AM · Jan 24, 2021 · Twitter Web App

Emily Burns ✓ DMs welcome #TeamReality
@Emily_Burns_V

What's really driving school closures. It's not COVID.

Look at these two maps, and tell me which pair looks similar.

Correlation is not causation, but when one party is trying desperately to instill fear in their populaces, that has its effect.

This is the effect.

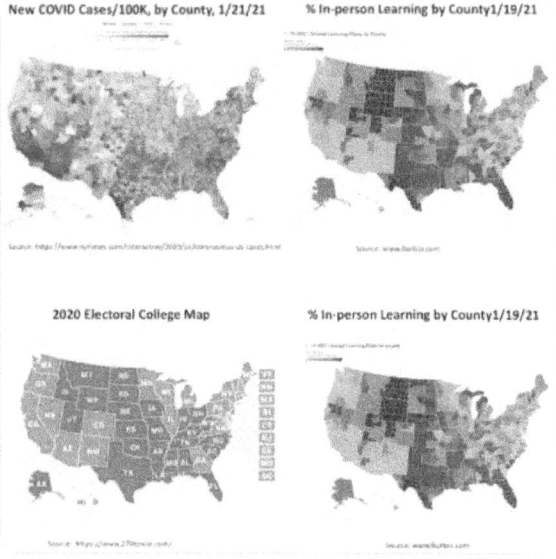

4:07 PM · Jan 22, 2021 · Twitter Web App

TEAM REALITY

 **Gummi Bear**
@gummibear737

If the US government keeps spending like they've been spending, the $15 minimum wage will be a pay cut in terms of purchasing power

Everything is stupid

2:54 PM · Jan 18, 2021 · Twitter for iPad

 **Clifton Duncan: Good Looking Loser.**
@cliftonaduncan

Cuomo has finally joined the club of selfish, grandma-killing, anti-science Right wingers that were shouted down for making this very same suggestion months ago

 **Archive: Governor Andrew Cuomo** @NYGovCuomo · Jan 11, 2021
We simply cannot stay closed until the vaccine hits critical mass. The cost is too high. We will have nothing left to open. We must reopen the economy, but we must do it smartly and safely.

#SOTS2021

5:36 PM · Jan 11, 2021 · Twitter for Android

Zac Bissonnette
@ZacBissonnette

It's been 10 months. At some point, you need to be able to show empirical evidence that these incredibly destructive covid restrictions work.

Not models. Not expert opinions.

The places that didn't do lockdowns should consistently have more deaths. And they just don't.

5:28 PM · Jan 29, 2021 · Twitter for iPhone

 **Emma Woodhouse**
@EWoodhouse7

🔴 STOP watching the news

🔴 START reading the data

10:50 PM · Jan 24, 2021 · Twitter for iPhone

 **Jenin Younes**
@Leftylockdowns1

If it was child abuse before 2020, it's still child abuse. So depriving kids of education, locking them in dorm rooms, isolating them, masking them- all child abuse, pandemic or no pandemic

8:45 PM · Jan 20, 2021 · Twitter for iPhone

 **Jennifer Sey**
@JenniferSey

A generation between the ages of 5-18 has been effectively removed from society at large. They do not have the same ability to vote or speak out, so it is time for children's advocates, incl teachers & parents, to raise their voices for a return to school.

washingtonpost.com/opinions/local...

10:33 PM · Jan 15, 2021 · Twitter for iPhone

FEBRUARY 2021

Prof
@covidtweets

Epidemiologists trying to understand why cases are falling.

Exhibits in the following tweets...

11:31 PM · Feb 6, 2021 · Twitter for Android

TEAM REALITY

 **Real Developments**
@pubdev

Mass hysteria in action...

Americans think Covid death rate is ~20%.

Pure insanity. No wonder people are so willing to lose all their freedoms.

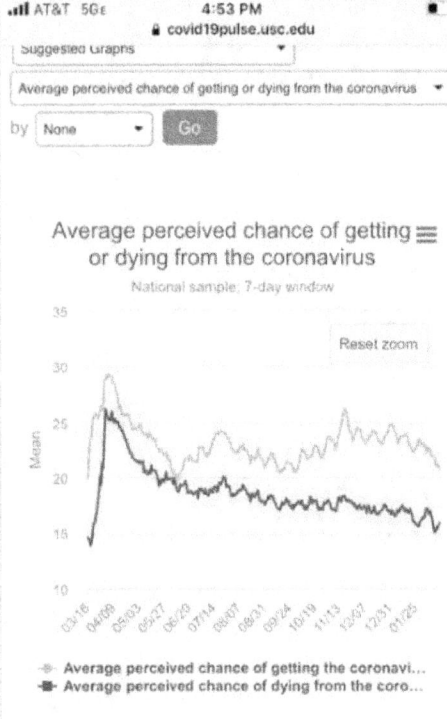

5:01 PM · Feb 18, 2021 · Twitter Web App

Emma Woodhouse
@EWoodhouse7

When is Fauci going to tell us to put a symbol on our front doors to show whether a residence is a House of the Vaccinated?

5:21 PM · Feb 23, 2021 · Twitter for iPhone

Infectious Disease Ethics
@ID_ethics

As one of my infectious disease physician colleagues said in March 2020:

If we could not produce immunity to novel pathogens, humans would be long extinct.

9:02 PM · Feb 23, 2021 · Twitter Web App

 **Hold2**
@Hold2LLC

I'd like to illustrate the absurdity of the new CDC guidelines.

Here is how the country looks based on the new color-codes zones (map and bar chart).

66.4% already offer in-person (40.8 5d/wk, 25.6% hybrid), yet CDC tells us only 9% should be open at all WITH all 5 mitigations.

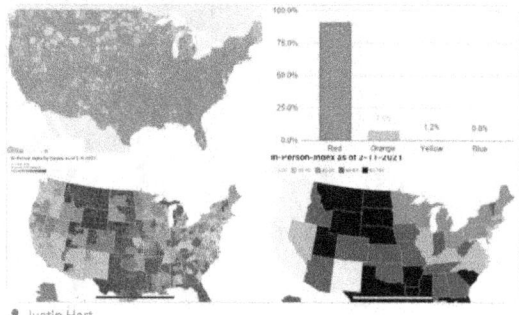

👤 Justin Hart

12:50 PM · Feb 14, 2021 · Twitter for iPhone

Kyle Lamb
@kylamb8

"There are some people more upset at me for vaccinating seniors than they are for other governors whose policies killed seniors. That is a joke"

-- @GovRonDeSantis
At presser in Brooksville

11:44 AM · Feb 24, 2021 · Twitter Web App

Ian Miller
@ianmSC

Let's check in on how our favorite experts are doing with their variant predictions of doom, shall we?

Yup...right on the money, as always

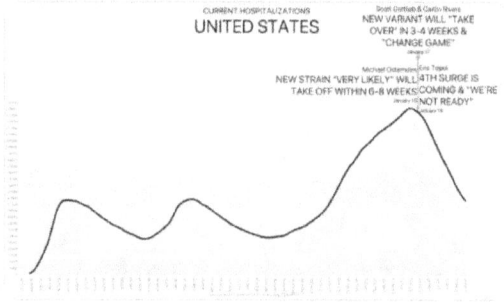

2:28 PM · Feb 24, 2021 · Twitter for Mac

TEAM REALITY

Brumby
@the_brumby

Ladies & gentlemen behold the null hypothesis: ~zero correlation between government response stringency (lockdown) and COVID mortality. The virus spreads until it doesn't. The burden is on the lockdowners to prove otherwise and they have failed.
1/2

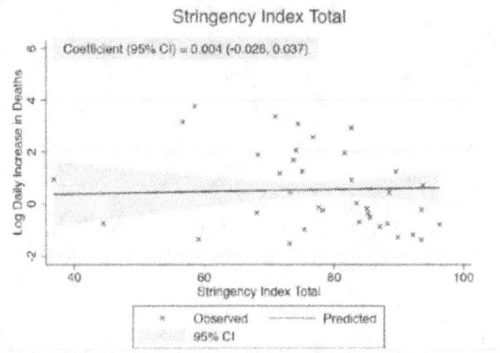

4:47 PM · Feb 4, 2021 · Twitter Web App

Josh Stevenson
@ifihadastick

Replying to @WHCOS

This is a total lie. Joe Biden's "50% of schools" goal is a complete farce. 50% of Districts were already open in November. And MORE than 50% of districts were ALREADY open the first week into his presidency.

Source:
cai.burbio.com/school-opening...

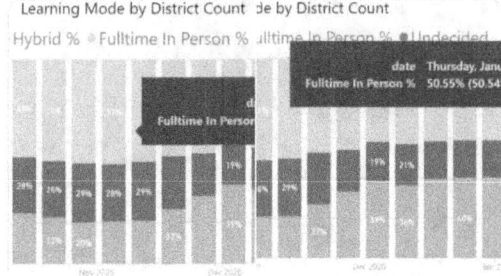

12:07 PM · Feb 18, 2021 · Twitter Web App

Phil Holloway™
@PhilHollowayEsq

I'm a pilot who was in a plane crash when I was young. A wise mentor told me to get back up ASAP or paralyzing fear could set in. I flew within an hour. The longer schools stay closed the harder it will be to avoid paralyzing fear. #OpenSchoolsNow or risk losing education forever

9:28 AM · Feb 7, 2021 · Twitter Web App

Josh Stevenson
@ifihadastick

Democrat States (by Governor) offer 18.7% (5.02M) of their Public School Students In-Person Instruction.

Republican States offer 65.59% (15.4M) of Public School Students In-Person Instruction.

11:56 PM · Feb 18, 2021 · Twitter Web App

John Ziegler
@Zigmanfreud

Please never forget that the core of a case for endless lockdowns is that we are supposedly protecting our seniors.

We are doing this by, in their last years of life, taking away at least 3 of their favorite things to do...

-Travel
-Lunch
-Visit grandkids

Makes perfect sense!

4:19 PM · Feb 1, 2021 · Twitter for iPhone

Megan "Hey it's the mask lady" Mansell
@mamasaurusMeg

Shamelessly stolen from Facebook.
Earning unrestricted breathing in schools. This is disgusting.

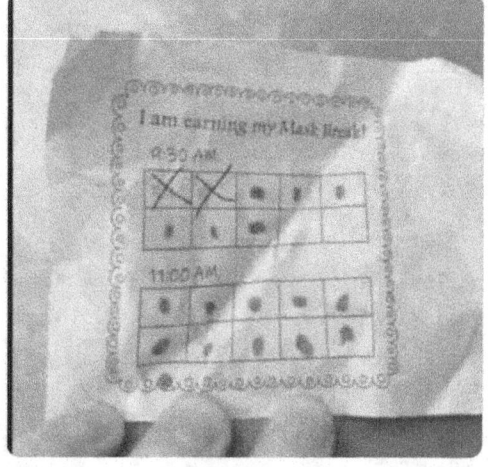

9:41 AM · Feb 25, 2021 · Twitter for iPhone

TEAM REALITY

Justin Hart @justin_hart

Seen in California.

9:26 PM · Feb 17, 2021 · Twitter for iPhone

Vanessa @vial42

My family is eating at a restaurant and our boys are outside playing tag with a group of kids they just met. Not everywhere in the US treats kids like biohazards.

8:15 PM · Feb 5, 2021 · Twitter for iPhone

Kevin McKernan @Kevin_McKernan

Evidence of rigged Peer Review
@PlanetWaves
@Bobby_Network
@RetractionWatch
@goddeketal

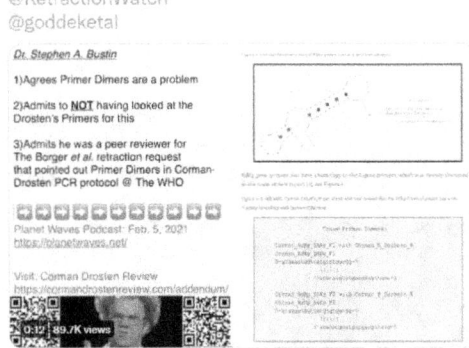

10:35 PM · Feb 9, 2021 · Twitter Web App

Vinay Prasad, MD MPH @VPrasadMDMPH

I am sorry. You can't make 2 million dollars in giving pandemic advice to for profit entities and run the government's pandemic response.

You just can't.

3:06 PM · Feb 20, 2021 · Twitter Web App

Phil Kerpen @kerpen

CDC confessed to letting teachers union overrule science. Outrageous corruption.

> **Fairfax County Parents Association** @FFXParentsAssoc · Feb 15, 2021
> 🚨 Breaking 🚨 @CDCDirector admits that lobbying groups changed recs in the recent CDC School Guidelines
>
> From a Fri. press conf: CDC based its recs on what the science says AND input from teachers groups w "direct changes in the guidance made as a result."
> cdc.gov/media/releases...
>
> Show this thread
>
> UNDERSTANDING OF THE LIVED EXPERIENCES, CHALLENGES, AND PERSPECTIVES, OF TEACHERS AND SCHOOL STAFF, PARENTS, AND STUDENTS.
>
> WE HAVE CONDUCTED AN IN-DEPTH REVIEW OF THE AVAILABLE SCIENCE AND EVIDENCE BASE TO GUIDE OUR RECOMMENDATIONS, AND WE HAVE ALSO ENGAGED WITH MANY EDUCATION AND PUBLIC-HEALTH PARTNERS, TO HEAR FIRSTHAND FROM PARENTS AND TEACHERS, DIRECTLY, ABOUT THEIR EXPERIENCES AND CONCERNS.
>
> PERSON INSTRUCTION IS NOT ONE THAT ANY OF US TAKE LIGHTLY. BELIEVE ME, I KNOW. AT CDC, WE HAVE THOROUGHLY REVIEWED THE SCIENCE, AND ENGAGED WITH STAKEHOLDERS AS WE WORKED TO PRODUCE AN OPERATIONAL STRATEGY, TO SUPPORT SAFE, IN-PERSON INSTRUCTION AND PROTECT TEACHERS, STUDENTS, AND OTHER SCHOOL STAFF.
>
> THANK YOU FOR YOUR ATTENTION. I LOOK FORWARD TO TAKING YOUR QUESTIONS. BUT FIRST, I WILL NOW TURN IT OVER TO DONNA HARRIS-AIKENS, FROM THE DEPARTMENT OF

8:14 AM · Feb 15, 2021 · Twitter Web App

Emily Burns   #SmilesMatter DM'... @Emily_Burns_V

1/many
Last weekend I escaped to Florida from Massachusetts, the fascist hellhole I am cursed to call home.

Everything you have heard is true. They have real people there, not the zombies that people the blue wastelands.

People smile, they laugh, they acknowledge you. Join me

7:28 PM · 4 Feb 2021

Justin Hart @justin_hart

2 weeks ago - Super Bowl! Superspreading event! Florida cases by day and Hillsborough County

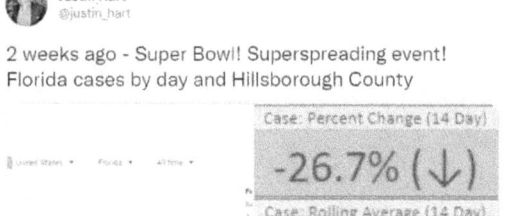
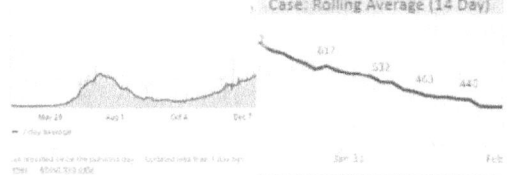

6:55 PM · Feb 19, 2021 · Twitter Web App

TEAM REALITY

Vinay Prasad, MD MPH
@VPrasadMDMPH

When experts go on television, they're incentivized to put the blame on average people: you didn't wear a mask, you didn't wear 2, you didn't stay home

They are strongly disincentivized for being critical of political leaders whose policies fail us

That is key to covid19

1:10 PM · Feb 25, 2021 · Twitter Web App

Wes Pegden
@WesPegden

The pandemic in a nutshell:

We have volumes of polling on whether Americans consistently wear masks "anytime they leave their house" but no regular surveillance on questions like "can you access paid sick leave", how many people fail to isolate when developing symptoms, and why.

10:22 AM · Feb 1, 2021 · Twitter Web App

Eric
@The_OtherET

A large RCT study on masks out of Denmark with 6,000 participants finds masks provide very little benefit, if any at all

Media: "Misinformation! This is dangerous!"

Two mannequin heads spray water at each other with two masks on

Media: "CONFIRMED! Double masks work!"

7:56 PM · Feb 19, 2021 · Twitter for iPhone

Woke Zombie
@AWokeZombie

#US #COVID19 If you want to process how insanely idiotic the Low Transmission Blue Category is

We would need to avg under 4600 cases a day across 7 days in the entire US.

We last did that on 3/19/20 when we reported 4,195 cases on 26k tests. Now we do 1.8M tests a day, 70x more

10:25 PM · Feb 14, 2021 · Twitter Web App

Eli Klein
@TheEliKlein

I have no problem with someone who had covid already, got vaccinated, and is double masked running alone outside

I have a big problem with that person shaping Covid policy

4:46 PM · Feb 25, 2021 from Manhattan, NY · Twitter for iPhone

---

Matt, Pre-School Diploma
@statomattic

Totally stupid virus-related thing that doesn't catch enough blowback:

It's been widespread in 50 states for months. Why are we still asking people if they've traveled out of state/making them isolate or test?

It's not as if by going to WI I could bring home some exotic plague.

10:08 AM · Feb 23, 2021 · Twitter for Android

Ann Bauer
@annbauerwriter

Pro-vax is very different from pro-mandate.

I'm also pro-marriage. Doesn't mean I think the government should be able to force you into one.

2:18 PM · Feb 21, 2021 · Twitter for Android

Ian Miller
@IanmSC

Well everyone. It's time.

It's officially been two weeks since the single worst thing that's ever happened to experts & media... maskless Super Bowl Celebrations

They've been wrong SO many times...surely THIS is the time a huge maskless gathering becomes a "superspreader"

Oof

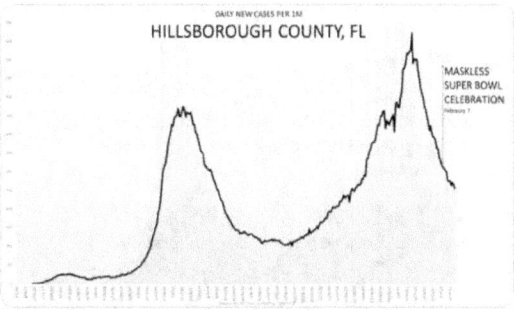

12:25 PM · Feb 21, 2021 · Twitter for Mac

TEAM REALITY

AJ Kay
@AJKayWriter

Keep in mind:

The pandemic didn't use erroneous data to model an inflated estimate of mortality.

The pandemic didn't write hyperbolic headlines.

The pandemic didn't create overly-inclusive reporting requirements.

The pandemic didn't design a test unfit for purpose...

1/5

11:37 AM · Feb 6, 2021 · Twitter for iPhone

The pandemic didn't quarantine the healthy. The pandemic didn't initiate lockdowns. The pandemic didn't decimate small businesses. The pandemic didn't close schools, some going on a full year. The pandemic didn't reduce the healthcare workforce... ++ The pandemic didn't ignore basic tenets of public health. The pandemic didn't exploit itself for political leverage. The pandemic didn't create a lucrative products/industries dependent upon its continuation. The pandemic didn't siphon resources from other deadly threats ++ The pandemic didn't use fear to control people's behavior. The pandemic didn't threaten people with social, financial, and professional ostratcization for pointing out the harms caused by any of the above. People did those things. They were not inevitable... ++ As we sort through the outcomes of the last year, it's critical that we trace each impact back to its source. Pretending like all of this destruction was inevitable does nothing to help us fix the damage now, & will do nothing to help us make better decisions in the future.

Prof Francois Balloux
@BallouxFrancois

I'm a big believer in Hanlon's razor ("Never attribute to malice that which is adequately explained by stupidity)".

But:

- Leak to journalist
- Release scary report full of holes
- Withhold data
- Rinse and repeat

... does not feel ideal in terms of Public Health messaging.

10:31 AM · Feb 2, 2021 · Twitter Web App

Hold2
@Hold2LLC

Remarkable.

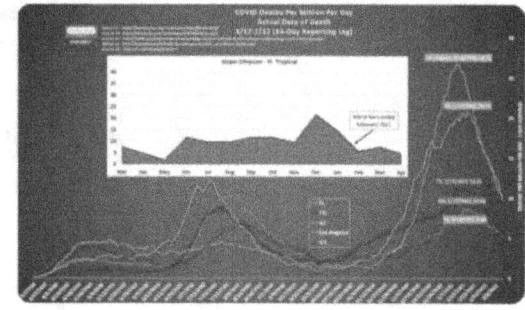

2:57 PM · Feb 28, 2021 · Twitter Web App

Matt. Pre-School Diploma
@statomattic

The main campus of my home state's state university is apparently now prohibiting students from exercising outdoors because science.

Not just in groups, not just mandating masks. You cannot go for a walk outside by yourself wearing three masks.

Science.

1:42 PM · Feb 11, 2021 · Twitter for Android

District AI
@districtai

I believe CDC's guidance today about masking was because they realize the data shows masking doesn't work so they have to lay blame on the type of masking or masking "technique". This is an admission that simply "face coverings" was a failed policy they pushed for almost a year

12:48 PM · Feb 10, 2021 · Twitter Web App

TEAM REALITY

Prof
@covidtweets

I am still waiting to see one convincing explanation for the current decline in cases that does not involve seasonality.

Most epis say people began taking more precautions. When I ask for any evidence or how everyone suddenly started behaving at the same time?

Crickets... (1/4)

9:36 AM · Feb 21, 2021 · Twitter Web App

... A decent number mention immunity build-up with many people now infected or vaccinated, which is definitely a factor, but still not enough by itself. That would mean prior infection rates are the same everywhere. Are you ready to admit that to be the case between GA and CA? ++ Seriously, I am desperately looking for one even remotely plausible explanation. If it is not seasonality, then what? Epis have great faith in that we are in control. But this is really nothing more than "faith" at this time, since the evidence is non-existent. ++ If anyone gives me convincing evidence about a significant behavior change occurring around early January, simultaneously everywhere, I promise that I will jump ship and be a fervent soldier of team "we should get the cases down". ++ ADDENDUM Re: #2: If infection rates are similar in GA and CA (and FL, etc.), it means around the same % got infected so far, despite widely different policies. Not sure those who make this claim are aware of this implication...

Stefan Baral ✓
@sdbaral

Worst public health response in history.

4:29 PM · Feb 19, 2021 · Twitter Web App

Mark Changizi
@MarkChangizi

"How dare you become so vocal about Covid hysteria after more than a decade of apolitical tweets?"

Maybe the fact that I am apolitical suggests there's something deeply wrong happening, and that it must be fought.

3:40 PM · Feb 7, 2021 · Twitter Web App

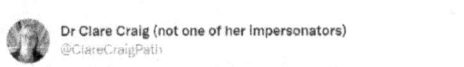
Dr Clare Craig (not one of her impersonators)
@ClareCraigPath

Winter respiratory viruses have one, or two, winter peaks. The fact every country has falling cases is wonderful news and evidence of endemicity. It is not evidence that the different interventions in each country simultaneously worked.

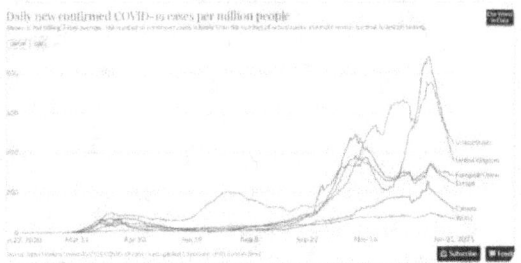

4:48 AM · Feb 2, 2021 · Twitter Web App

Emma Woodhouse
@EWoodhouse7

I don't care what you call it.

Social contagion
OCD
Mass delusion event
Mass hysteria
Stockholm Syndrome
Munchausen Syndrome [by proxy or not]
All of the above

It's happening, & it needs to END.

11:41 PM · Feb 23, 2021 · Twitter Web App

TEAM REALITY

District AI
@districtai

I cannot stress this enough. This drop we have experienced is not driven by behavior. We have decent data from IHME and USC behavior that shows it isnt. It is combination of seasonal stimulus and herd immunity/resistance.

9:29 PM · Feb 25, 2021 · Twitter Web App

Matt, Pre-School Diploma
@statomattic

If you get a coronavirus shot and don't post a photo to brag about it on social media, are you actually vaccinated?

1:06 PM · Feb 24, 2021 · Twitter for Android

Bethany S. Mandel
@bethanyshondark

Remember when we used to acknowledge how unhealthy extended amounts of screen time is for children?

4:25 PM · Feb 22, 2021 · Twitter for iPhone

Gummi Bear
@gummibear737

Before Covid-19, this was the CDC playbook for Influenza Pandemic Non-Pharmaceutical-Interventions (NPIs)

Notice how even in the worst case (think Spanish Flu) they don't recommend:
-Universal masking
-Business shutdowns
-Lockdowns

Everything they're doing is CCP pseudoscience

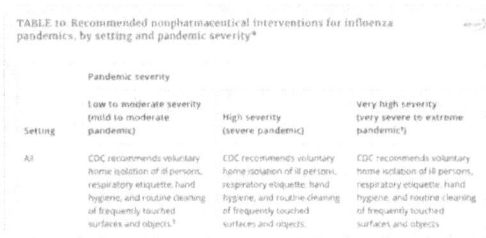

4:28 PM · Feb 10, 2021 · Twitter for iPad

Eric
@The_OtherET

A person at low risk for COVID isn't performing some righteous act by staying home all day, taking Zoom meetings and ordering necessities through Amazon and Doordash. They're just pushing the economic and viral burden onto the people that can't do these things.

12:07 PM · Feb 11, 2021 · Twitter for iPhone

Emma Woodhouse
@EWoodhouse7

SWEDEN 🇸🇪: No excess deaths under age 75 in 2020.

UNITED STATES 🇺🇸: Lots of excess deaths under age 75, including tens of thousands under age 50 that had nothing to do with "that" virus.

I choose Sweden.

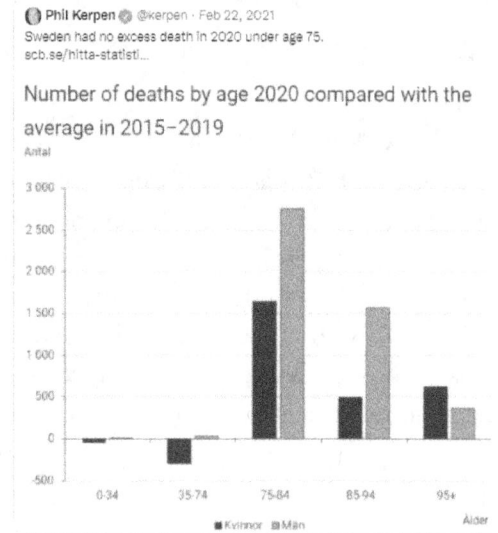

11:08 AM · Feb 22, 2021 · Twitter Web App

Emily Burns DMs welcome #TeamReality
@Emily_Burns_V

1/n
Repressive COVID policies are NOT tied to a reduction in deaths.

Repressive COVID policies ARE linked to EXTREMELY low levels of in-person learning, and high unemployment.

#ZeroCOVID basically equates to #ZeroSchool, #ZeroJobs, #ZeroLife--and the same death.

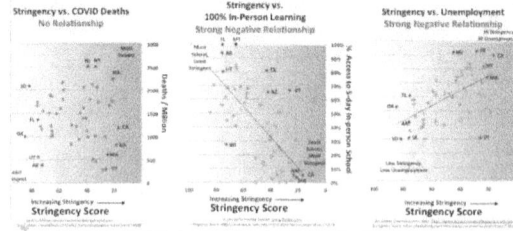

3:55 PM · Feb 17, 2021 · Twitter Web App

Prof
@covidtweets

Seasonality thread...

Our "experts" had no idea why cases were falling. Now, they have no idea why they are rising.

Because they cannot accept how strong seasonality is. They like to think their advice is making a difference. (1/x)

> **Prof** @covidtweets · Feb 6, 2021
> Epidemiologists trying to understand why cases are falling.
>
> Exhibits in the following tweets...
> Show this thread

I have long predicted that we would see a rise in March. I will explain in the following tweets why I thought that would happen. TL;DR: It has nothing to do with the variants or opening up too soon. ++ First, I came across a study last year, which describes a specific temp profile in which the cases were exploding. It found that the virus didn't spread effectively in very cold weather. It fit my observations as well. The critical range: 41-52 F. ++ Based on this, I predicted that winter would suppress the virus in most places, except for the South. My poor choice of words make it look like I was wrong, but cases actually declined throughout winter in more than half of the states. Until now... ++ If the critical range is 41-52, that would mean two waves: The first one is when passing through that range downward, and second is when passing through it upward. When I was speculating this, the "experts" were waiting for the Winter apocalypse. ++ How about the Sunbelt wave, I am often asked. I long believed that there were actually two ranges, not one, based on another study I had seen earlier (cannot find it now). But another study confirmed that, finding there is another, warmer bracket. ++ Seasonality is not *just* weather, but I actually think the virus only explodes when it can go airborne. And it can only go airborne under very specific conditions. The two studies above provide evidence for this. ++ When temp/humidity is not favorable, main route is droplets, so spread actually slows down. But for a period of time (maybe as short as 2-3 weeks), the virus effectively goes airborne (likely incl. outdoors). That's likely what is causing sharp spikes like the one below. ++ After a period of explosive spread, the infected population becomes enough to sustain a certain level of cases for a while, causing the long tails we usually see. Interestingly, the shape of curves in Summer are less steep, suggesting that bracket may be less favorable. ++ Now, what should we expect? It increasingly looks like this will be very similar to the 2020 March-April wave. It will be mostly localized in the North. I expect the peak to happen around April 5th. I expect NY to peak at around 13-15K cases 7-day average. ++ However, national numbers will likely be lower than the December numbers, since the most populous states will likely not be joining this wave. Also, the link between cases/hosp/deaths will be different, with those numbers hopefully staying lower with vaccines picking up.

TEAM REALITY

Emily Burns 🧬 DMs welcome #TeamReality
@Emily_Burns_V

17/n
To visualize this, you have to look at deaths on a population basis, BY AGE GROUP

Doing so, excess deaths are virtually imperceptible <65. In FL, even above 65, deaths are not much higher than prior flu years. In MA deaths begin to skyrocket THE DAY lockdowns start.

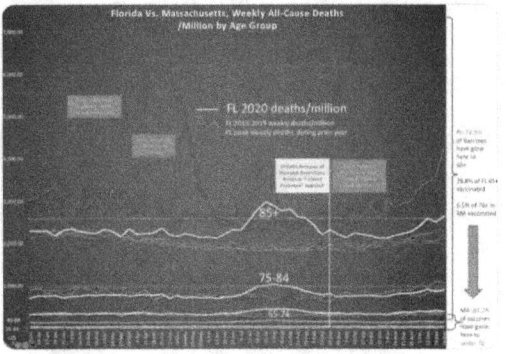

10:55 PM · Feb 4, 2021 · Twitter Web App

Emily Burns 🧬 DMs welcome #TeamReality
@Emily_Burns_V

1/many

COVID has been politicized. Children and families in blue states are paying the price. On average, in red states, 3x as many children have access to 100%, 5-day/week in-person learning as in blue. Nearly 4x as many children in blue states are 100% remote (sources @ end)

Age	5-day/week In-Person Learning (% kids w/access)	Average Stringency (1=more stringent)	100% V (% of Ch
02	20%	35	44
34	62.6%	64	12.0

7:47 PM · Feb 9, 2021 · Twitter Web App

Eric
@The_OtherET

Eric Topol "B.1.1.7 beast" projection update

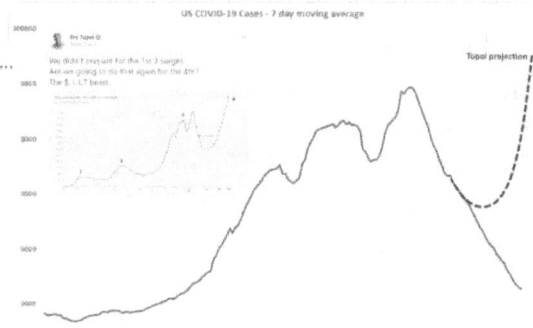

4:4 PM · Feb 22, 2021 · Twitter Web App

Eric
@The_OtherET

Fauci warned of a "surge upon a surge" of #COVID19 hospital patients from holiday gatherings

Here are 15 states with 15 different COVID policies affecting nearly 80 million people

Patients peaked well before Christmas in every single state - six of them even before Thanksgiving

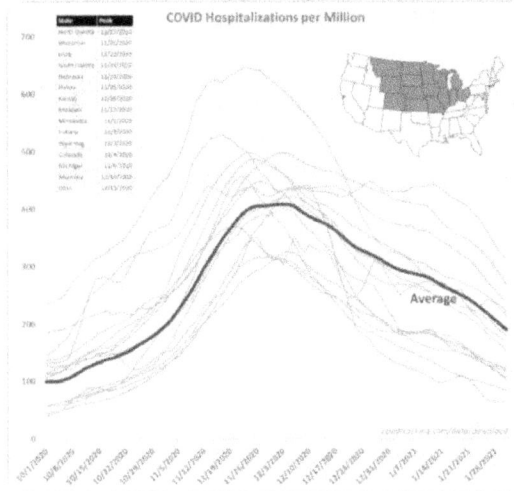

5:54 PM · Feb 1, 2021 · Twitter Web App

ClownBasket 🤡
@ClownBasket

5 Mask Questions
-If smells get thru my mask, can SC2?
-Would today's masks work against WWI mustard gas?
-Why did it take till 2020 to discover that masks stop viruses?
-If 2 masks are better, why not improve the 1 mask?
-Why were so many 1918 Flu deaths frm bacterial pneumonia?

7:05 PM · Feb 27, 2021 · Twitter for iPhone

MARCH 2021

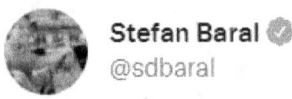

Stefan Baral ✓
@sdbaral

Ok, I found another #COVID19 hill that I am ready to die on....vaccine passports.

A policy so obviously problematic that only HYDRA could have conceived of it.

10:17 PM · Mar 25, 2021 · Twitter Web App

TEAM REALITY

 Ian Miller @ianmSC · Mar 29, 2021

I'm legitimately unsure if Biden's aware of what's going on at this point, but let's look at what's happened in what he described as the "Neanderthal" states that lifted mask mandates vs Michigan, New Jersey & New York, which all follow science & kept their mandates.

Whoops!

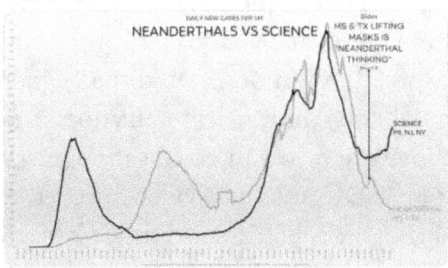

 Federico Andres Lois @federicoiois

I live in a country that has mandated masks even in outdoors (where I don't use it, period). I vote with my wallet, BUT, for maximum civil disobedience this is my mask for when I have no choice (shops, kids school, etc). It reads: "Doesn't filter aerosols. Doesn't work for CV19"

9:34 PM · Mar 6, 2021 · Twitter Web App

90 Retweets 7 Quote Tweets 378 Likes

 Stinson Norwood @snorman1776

"Starting now, everyone should try to avoid going to the ER."

Words from @DrLeanaWen's op-ed in the Washington Post.

They did stay home. They died.

9:57 AM · Mar 13, 2021 · Twitter Web App

 Vinay Prasad, MD MPH @VPrasadMDMPH

If your pandemic plan is that 1 year into a pandemic with nearly no resources provided to vulnerable people, you reiterate a fear based message and ask avg folks to find a fourth wind for voluntary compliance with restrictions

I would say your pandemic plan is not that good

5:30 PM · Mar 29, 2021 · Twitter Web App

 **Hold2** @Hold2LLC

FL vs. FL Mask Comparison: 3/25/21

Here are the current Cases/100k by county based on:

Red: Masks required in classrooms

Green: Masks NOT required while children seated in classroom

Yellow: Mask requirements vary by grade and/or individual schools

*Data current as of 3/24

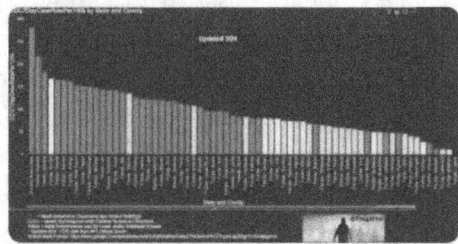

🧟 Woke Zomble

6:10 PM · Mar 25, 2021 · Twitter Web App

 Infectious Disease Ethics @ID_ethics

Fearful claim that #covid19 variants are likely to evolve to be more harmful to children

If so, why did this never happen with other coronaviruses?

Covids OC43, NL63, 229E (etc) just go on causing the common cold (+ rarely other effects) in kids, generation after generation...

> **New York Times Opinion** @nytopinion · Mar 30, 2021
> Covid "variants that cause worse disease among children are likely to emerge from children themselves, especially with adults becoming less hospitable hosts for infection as vaccinations rise," write @jeremyfaust and @angie_rasmussen. nyti.ms/3foGsnr

8:00 PM · Mar 30, 2021 · Twitter Web App

 Kyle Lamb @kylamb8

We're two days away from one year since 15 days to flatten the curve.

11:44 AM · Mar 14, 2021 · Twitter Web App

 Mark Changizi @MarkChangizi

The greatest disappointment over the last year wasn't Bill Gates, the WHO, politicians, academics or journalists, but, instead, our neighbors, friends and family.

4:57 PM · Mar 1, 2021 · Twitter Web App

TEAM REALITY

Tracy Hoeg, MD, PhD
@TracyBethHoeg

This morning, I testified as an expert physician epidemiologist on COVID in schools for a House of Representatives Round Table. It was a fascinating and productive experience. To me, the most important messages were as follows:

6:39 PM · Mar 12, 2021 · Twitter Web App

1. It is counter-productive to blame school districts/teachers for prolonged school closures when the message we are getting from the CDC is that schools are not safe at X amount of community spread/X amount of distance between students/until strict ventilation standards are met ++ We scientists know none of the above are based in K-12 school reopening science. We need a unified message from our CDC that is evidence-based which facilitates opening asap with simple, straight-forward, achievable mitigation strategies. ++ 2. The message that schools are overall safer than the community for kids both in terms of COVID and overall health and safety needs to get out and can't be emphasized enough. ++ (teachers have also been found to be at lower risk of severe disease than other essential workers and now that most are vaccinated, are very safe at work) ++ 3. Millions/billions of $ should not be spent on new school ventilation, expansions, plastic barriers when we know these are not necessary. This $ should instead be directly funneled to improve education and accelerate learning to make up for losses incurred over the past year. ++ 4. Prioritizing our children's education and well-being is something that should have united our country, but instead divided us. It should have been a top if not *the* top priority from the beginning to get our kids back in schools. ++ We adults should not have let politics interfere with this shared goal. When we prioritize the well-being of our kids, everything else falls into place. And happy 70th birthday to my mom, who has instilled in me my whole life: together we are stronger.

John Ziegler
@Zigmanfreud

In a rational world, because young/healthy people are not vulnerable to COVID, sports should be leading us BACK to normalcy.

Instead, because fear/wokeness has taken them over, they are 1 of the most powerful forces keeping us in the quicksand, ironically to their own detriment.

10:57 AM · Mar 21, 2021 · Twitter for iPhone

Vinay Prasad, MD MPH
@VPrasadMDMPH

The most regressive, harmful, racist, unjust, illogical, unfounded, absurd intervention to slow the spread of sars-cov-2 has been school closure, particularly of elementary school kids.

Even after we knew better, and all of Western Europe showed us how, we still made excuses

2:50 PM · Mar 25, 2021 · Twitter Web App

TEAM REALITY

Victoria Fox
@drvictoriafox

Pre 2020 Lockdowns were considered to be too destructive to implement & ineffective at controlling disease long term. In 2020 they were presented as a last resort SHORT TERM measure & no other choice to prevent extinction level deaths & hospital meltdowns. 1/5

12:53 PM · Mar 25, 2021 · Twitter for iPhone

We now know from several states/cities/countries that COVID outbreaks can be managed without lockdowns. There are plenty of models for keeping society & health care functioning that result in similar mortality rates to lockdown cities/nations. 2/5 ++

It's time to move to less destructive, supportive, resource-based management strategies in countries that can't eliminate COVID & keep their borders permanently closed (most countries in the world). ++ 3/5 Strategies that prioritize those most at risk & use sustainable PH measures, deliverable within a localities infrastructure that balance total harms. Equitable Measures that match the cultural/structural make up of society that people can/will comply with. 4/5 ++ PROLONGED use of lockdown over multiple months/years without a respite is destructive & counterintuitive if COVID continues to be widespread. It will take years to fully understand the harms caused by prolonged disruption to education, business, employment & Heath care 5/5

Eric
@The_OtherET

Took a weekend Twitter break and came back to see Fauci is saying kids can't play together without masks until they're vaccinated and Twitter censoring a Harvard epidemiologist for saying not everyone needs to get a COVID vaccine, so clearly everything is still going just great

6:48 PM · Mar 28, 2021 · Twitter for iPhone

Craig
@TheLawyerCraig

Today's #Covid19 Update Thread
(data pulled directly from state dashboards)

All graphics contain "7-day averages" for the metrics, but I'll post the raw counts daily in the second post of the thread as well.

UNITED STATES

UNITED STATES COVID19 Data as of March 8 *All Numbers are 7-Day Rolling Averages*								
Daily Tests		% chg	**Daily Cases**		% chg	**% Positive**		% chg
Today	1,433,139	-	Today	56,857	-	Today	3.97%	-
1 wk ago	1,511,435	-5%	1 wk ago	66,436	-14%	1 wk ago	4.40%	-10%
2 wks ago	1,331,281	+8%	2 wks ago	68,820	-17%	2 wks ago	4.79%	-17%
1 mo ago	1,680,850	-15%	1 mo ago	125,116	-55%	1 mo ago	7.44%	-47%
2 mos ago	1,662,378	-14%	2 mos ago	220,415	-74%	2 mos ago	13.26%	-70%
Peak (1/13)	2,033,495	-30%	Peak (1/11)	247,111	-77%	Peak (4/11)	20.45%	-81%
Daily Deaths		% chg	**In Hospital**		% chg	**In ICU**		% chg
Today	1,547	-	Today	42,663	-	Today	8,684	-
1 wk ago	1,892	-18%	1 wk ago	50,656	-16%	1 wk ago	10,412	-17%
2 wks ago	1,916	-19%	2 wks ago	59,756	-29%	2 wks ago	12,484	-30%
1 mo ago	2,996	-48%	1 mo ago	91,916	-54%	1 mo ago	18,291	-53%
2 mos ago	2,656	-42%	2 mos ago	126,929	-66%	2 mos ago	23,264	-63%
Peak (1/13)	3,335	-54%	Peak (1/12)	131,127	-67%	Peak (1/14)	23,771	-63%

6:46 PM · Mar 8, 2021 · Twitter Web App

Craig
@TheLawyerCraig

The regional trends in #COVID19 metrics seem to pretty clearly have a geographical or climatological tie. The states in RED have the largest recent percentage increase in positive testing %. The states in GREEN have the largest percentage decrease.

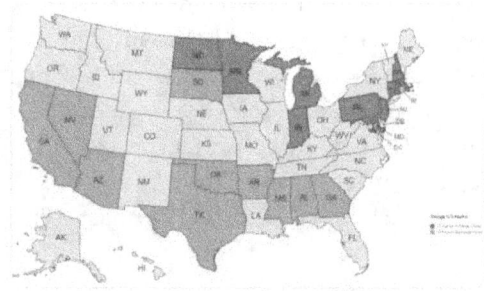

2:54 PM · Mar 26, 2021 · Twitter Web App

District Al
@districtal

The CDC has put out two flawed studies on mask mandates (temporal issues). Today put out one of the worst studies I've ever seen and has used mannequin lab experiments to try to show the effectiveness of masks. All show how desperate the CDC is. All very bad science.

1:48 PM · Mar 5, 2021 · Twitter Web App

Vanessa
@vial42

When the dust settles, we will need to seriously evaluate public health's ability to dictate every aspect of our lives. More importantly, we may need to codify into law a children's bill of rights to ensure our adult-centric society never does this to the kids again.

6:56 PM · Mar 24, 2021 · Twitter for iPhone

Brumby
@the_brumby

1 yr ago today, Dr. Ioannidis wrote the below. I read it that day, agreed with nearly every word, and still do. Worth revisiting in full, but I've captured a few key quotes below along with my own commentary

My favorite, which sums up the last yr:
1/8

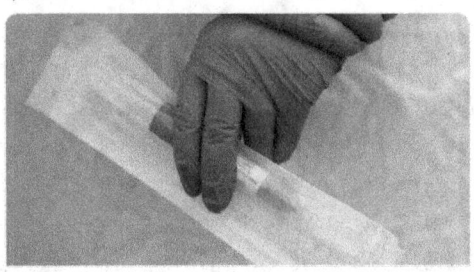

statnews.com
A fiasco in the making? As the coronavirus pandemic takes hold, we are makin...
A fiasco in the making? As the #coronavirus pandemic takes hold, we are making decisions without reliable data.

10:12 AM · Mar 17, 2021 · Twitter Web App

"we dont know how long social distancing measures and lockdowns can be maintained without major consequences to the economy, society, and mental health. Unpredictable evolutions may ensue, including financial crisis, unrest, civil strife, war, and a meltdown of the social fabric" ++ "How long, though, should measures like these be continued if the pandemic churns across the globe unabated" I'd love to know. We never should have implemented such devastating & unprecedented "temporary" measures without defined exit criteria. ++ "How can policymakers tell if they are doing more harm than good" Good question. See CA/FL. On lockdowns - "totally irrational. It's like an elephant being attacked by a house cat. Frustrated and trying to avoid the cat, the elephant accidentally jumps off a cliff and dies." ++ On deaths-"in an autopsy series that tested for resp viruses in specimens…who died during the 2016 to 2017 influenza season…respiratory virus was found in 47%." "A positive test for coronavirus does not mean…[the virus] is primarily responsible for a patient's demise" ++ On the possibility of long lockdown- "life largely stops, short-term and long-term consequences are entirely unknown, and billions, not just millions, of lives may be eventually at stake" Right; as I've said there's a reason no one has ever attempted to halt society before ++ "If we decide to jump off the cliff, we need some data to inform us about the rationale of such an action and the chances of landing somewhere safe" Agreed. I'm still waiting for someone to show the cost/benefit analysis they undertook prior to a decision to lockdown. ++ When the history books are written on our COVID response, they will spell vindication for Dr. Ioannidis & others brave enough to fight the mass hysteria. For the next decade, replace the word "pandemic" with "lockdown" in every article you read about the disaster created. 8/8

Matt, Pre-School Diploma
@statomattic

Just heard on CBS radio news:

"COVID cases are dropping across the nation, but doctors are alarmed that the testing rate is also dropping."

Stick with me here.. is there ANY chance that maybe fewer people are getting tested b/c fewer are getting sick?

The media is shameless.

9:10 AM · Mar 1, 2021 · Twitter for Android

Michael P Senger
@MichaelPSenger

Calling it now: in 5 years nearly everyone will tell you they knew lockdowns were a fraud from day 1.

9:53 AM · Mar 29, 2021 · Twitter for iPhone

Real Developments
@pdubdev

Kids should be zero feet apart.

Right on top of each

Playing and getting dirty

Social distancing is bullshit

8:05 PM · Mar 19, 2021 · Twitter Web App

TEAM REALITY

Show Me The Data
@txsalth2o

My friend's sister died of CV-19 last week.

She was 36.

She was also roughly 260 lbs.

Can you imagine what the world would be like if the government had been honest about who CV19 kills and addressed those issues?

4:20 AM · Mar 16, 2021 · Twitter for iPhone

Letting the obese die is not kindness. You are not a "good person" for not wanting to draw attention to that 78% of all CV19 hospitalizations are overweight or obese. Govmnt should have focused wellness, health and activity but instead enacted policies exacerbated the problem. ++ Just like ignoring the media continued to ignore nursing homes while focusing on beaches- they're ignoring the deaths of the obese while focusing on schools, where not a single death has been traced to pediatric transmission of CV19. ++ What if instead of "Teacher Dies of CV19" stories that were written specifically so the reader would draw an incorrect correlation between the dead's career and cause of mortality- they wrote "obese person dies of CV19" where there is an actual correlation. ++ We've known for 11 months that obese people are at a substantially higher risk. How much weight could my sister's friend have lost in 11 months? Our media and our government are complicit in her death. The fear of fat shaming is literally killing the obese. ++ There's a number of people pushing back on this thread with the same argument- that the obese CV19 hospitalization rate mirrors the obese CV19 population in the US so it's correlation not causation. I'm going to save you some time. [Link shared: https://consumer.healthday.com/3-3-covid-death-rates-10-times-higher-in-countries-where-most-are-overweight-report-2650888507.html]

Martin Kulldorff
@MartinKulldorff

After having protecting themselves while the working class were exposed to the virus, the vaccinated #Zoomers now want #VaccinePassports, where immunity from prior infection does not count, despite stronger evidence for protection. One more assault on working people.

3:04 PM · Mar 17, 2021 · Twitter Web App

Eli Klein
@TheEliKlein

~60% of Americans think the chances somebody with Covid must be hospitalized are ~10x higher than they actually are

Survey by Gallup and Franklin Templeton

6:56 PM · Mar 20, 2021 from Manhattan, NY · Twitter for iPhone

Prof
@covidtweets

I wanted to check to see how Neanderthal states were doing compared to homo sapiens.

Shout out to @TheLawyerCraig for the data.

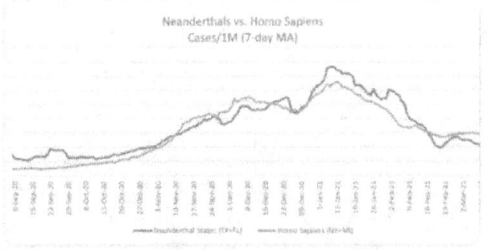

10:48 PM · Mar 18, 2021 · Twitter Web App

AJ Kay
@AJKayWriter

If, after a year, we still "don't know enough" about #COVID19 to manage it without shutting down the world,

Then the so-called 'experts' who are supposed to be figuring it out should be fired and replaced.

Doing this poor of a job in any other profession wouldn't fly.

12:25 AM · Mar 15, 2021 · Twitter for iPhone

TEAM REALITY

Emily Burns DMs welcome #TeamReality
@Emily_Burns_V

4/n
The facileness of this thinking is almost breathtaking... They don't work for flu or SARS-CoV1, but just MIGHT work for SARS-CoV-2, b/c Asian countries are "maskier" ?

Something about this virus, makes it more likely to avoid the laws of physics than other viruses?

4:22 PM · Mar 4, 2021 · Twitter Web App

Martin Kulldorff
@MartinKulldorff

Anti-vaxxers and fanatics pushing mandatory vaccinations and vaccine passports are equally bad for public health. After the lockdown and contact tracing fiascos, the latter generate vaccine skepticism. Only by trusting people can we restore peoples trust in public health.

> **Martin Kulldorff** @MartinKulldorff · Dec 19, 2020
> #9 Public health is about trust. To gain the trust of the public, public health officials and the media must be honest and trust the public. Shaming and fear should never be used in a pandemic.
> thehill.com/opinion/health... @camakridis
> Show this thread

8:44 AM · Mar 11, 2021 · Twitter Web App

Prof Francois Balloux
@BallouxFrancois

I don't doubt masks reduce transmission of #SARSCoV2, based on basic physical principles, even though their efficacy may be marginal in real life settings. Though, their psychological impact isn't subtle, as I had anticipated nearly a year ago.

> **Prof Francois Balloux** @BallouxFrancois · Mar 24, 2020
> There is no doubt that masks reduce Covid-19 transmission, as do other behavioural changes. A more relevant question to me is to what extent individuals in different societies would welcome radically altering their way of life, long-term. I predict difficult discussions ahead. twitter.com/SteveStuWill/s...

5:55 PM · Mar 2, 2021 · Twitter Web App

Martin Kulldorff
@MartinKulldorff

When making unscientific claims, media often refer to "health officials" or "experts" without citing anyone. I challenge @Twitter to name vaccine experts believing that everyone, including children and those with prior immunity from natural infection, MUST be Covid vaccinated.

> **Martin Kulldorff** @MartinKulldorff · Mar 15
> No. Thinking that everyone must be vaccinated is as scientifically flawed as thinking that nobody should. COVID vaccines are important for older high-risk people and their care-takers. Those with prior natural infection do not need it. Not children.
> ⓘ This Tweet is misleading. Learn why health officials recommend a vaccine for most people.

5:57 PM · Mar 28, 2021 · Twitter Web App

Woke Zombie
@AWokeZombie

#FL #COVID19 Masked mandated school districts shown in Red. School districts without mask mandates are shown in Green. Data as of 3/12, sourced from NYT Github database.

Mask mandated counties not holding up too well here.

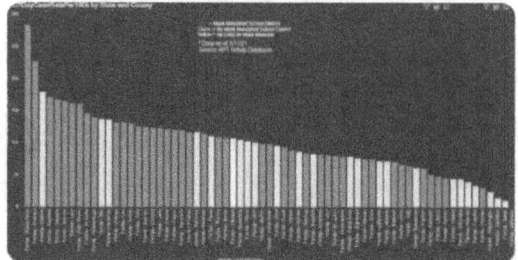

5:32 PM · Mar 12, 2021 · Twitter Web App

Emily Burns DMs welcome #TeamReality
@Emily_Burns_V

1/n
We are in a 5-alarm social justice emergency. But the culprits are those states whose populations claim to care the most about social justice. BLM support is tightly tied to repressive COVID policies that result in stunningly low access to education and high unemployment.

State	BLM Support	% 100% Virtual	% Kids w/5-day Traditional In-person	Stringency Rank (1= Most Stringent)	Deaths/M Rank (least to most)	Unemployment Rate %
VERMONT	66%	8%	67%	7	2	3.1
MASSACHUSETTS	64%	42%	4%	3	49	7.4
MARYLAND	63%	84%	0%	28	22	6.3
CALIFORNIA	60%	44%	5%	1	19	9
HAWAII	59%	54%	0%	9	1	9.3
RHODE ISLAND	59%	19%	42%	18	43	8.1
WASHINGTON	58%	66%	2%	6	7	7.1
NEW YORK	57%	74%	24%	5	50	8.2
OREGON	57%	24%	6%	15	5	6.4
CONNECTICUT	55%	11%	28%	11	45	8
ILLINOIS	54%	36%	11%	10	38	7.6
MAINE	54%	0%	0%	13	4	4.9
NEW HAMPSHIRE	54%	3%	13%	29	9	4
NEW JERSEY	54%	39%	3%	16	51	7.6
COLORADO	52%	0%	54%	14	14	6.4

3:32 PM · Mar 2, 2021 · Twitter Web App

Jenin Younes
@Leftylockdowns1

Can we please remove the word 'safe' from the English lexicon? It's now like nails on a chalkboard to me

6:15 PM · Mar 21, 2021 · Twitter for iPhone

Ann Bauer
@annbauerwriter

My son, an essential worker who had Covid last spring, coughed once at his new job yesterday and was immediately discharged, told to file for unemployment and get 2 neg Cov tests b4 returning.

Because in MN we have tons of tax money to pay people! And we SCIENCE all the time!!

2:54 PM · Mar 24, 2021 · Twitter Web App

TEAM REALITY

Gummi Bear
@gummibear737

After one year of lockdown, my blood pressure has gone from being perfect all my life to elevated (I'm 44)

Weight gain was minimal (10 lbs)

But sedentary lifestyle...

Inactivity due to lockdown is a global killer

Effects will not be felt for years

We've abandoned science

4:19 PM · Mar 3, 2021 · Twitter for iPad

Emma Woodhouse
@EWoodhouse7

Wear a mask for 30 days!
Wear a mask for 30 more days!
Wear a mask for 100 days!
Wear a mask until vaccine!
Wear a mask after vaccine!
Wear a mask forever!

(See how this works?)

12:43 PM · Mar 22, 2021 · Twitter for iPhone

Brumby
@the_brumby

If COVID hit just 10yrs ago "lockdowns" wouldve never been entertained b/c elites couldn't Zoom at home & have DoorDash deliver kale to their Peloton while the poor lost their jobs & absorbed the viral burden. Tech didn't help us weather a crisis; tech perpetuated the real crisis

Age-Adjusted Death Rates due to COVID-19 per 100K March 22, 2021			
			Mortality Rate
Los Angeles County Total			209
Race/Ethnicity	Asian		142
	Black/African American		187
	Hispanic/Latino		337
	White		118
Area Poverty	<10% area poverty		118
	10% to <20% area poverty		213
	20% to <30% area poverty		288
	30% to 100% area poverty		393

7:22 PM · Mar 26, 2021 · Twitter Web App

Emma Woodhouse
@EWoodhouse7

It's incredible - and heartbreaking - that the first 8 months of 2020 saw a 14,000+ increase in Drug Overdose death over the prior 3Y avg.

14,000. Let that sink in. #EpicPolicyFails

data.cdc.gov/NCHS/Monthly-C...

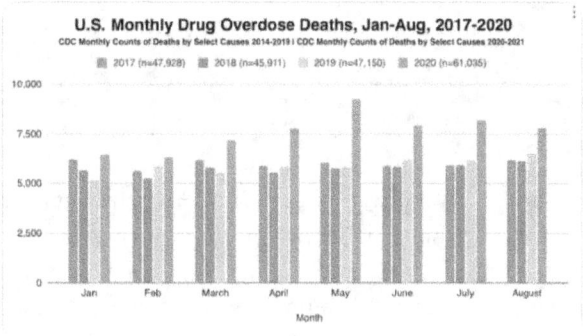

11:33 AM · Mar 27, 2021 · Twitter Web App

Emily Burns DMs welcome #TeamReality
@Emily_Burns_V

1/
Lockdown-driven public health interventions have skewed COVID-19 infections in a deadly way, driving the virus away from the young, and towards the old. This distribution looks nothing like other pandemics over the last 100 years.

academic.oup.com/epirev/article...

Age-Based Infection Attack Rate: COVID-19 vs. 1957 Flu (U.S. Only)				
Age Group	% Infected w/COVID-19 in US	% Infected in 1957 flu	COVID-19 Infection Fatality Rate**	Incremental Deaths from Unnatural Infection Distribution Through January 9, 2021
<5	15%	32%	0.002%	(72)
5-17	27%	44%	0.001%	(34)
18-49	30%	26%	0.031%	2,082
50-64	23%	17%	0.330%	11,919
65+	17%	8%	2.843%	138,482
Total/Overall	25%	24%	0.38%	152,377

7:31 PM · Mar 18, 2021 · Twitter Web App

Emily Burns ✓ DMs welcome #TeamReality
@Emily_Burns_V

1/

Masks ARE political. They have no basis in science—except perhaps political science—and have ZERO impact on COVID deaths or cases.

Pitched as no-cost interventions they are in-truth linked to incredible societal harms--visited in proportion to their adoption.

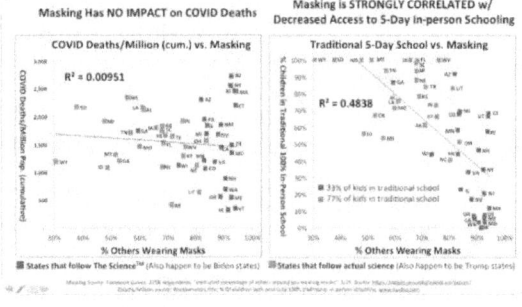

9:04 PM · Mar 30, 2021 · Twitter Web App

While there is no linkage between masking and COVID deaths, they are linked to ultra-low levels of in-person schooling. If these two charts were flipped, masks would be a no-brainer. That they are not means masks—and the fear they generate—are causing grave harm. ++ Masks are also linked to significantly higher levels of unemployment—driving closures in services industries, and generally making people afraid of engaging with each other. The brunt of this is borne by blue states with strong mask mandates. ++ The harms caused by masks—low-levels of education, high unemployment—are felt most keenly by minorities. Yet, by far the highest correlation with % masking is support for BLM. People are unwittingly betraying their highest principals, due to a political perversion of science. ++ While masks and BLM are tightly linked, masks do not have any impact on whether or not states have higher percent of deaths of minorities relative to their populations ++ In fact, comparing Florida and California, when restrictions were lifted in Florida—masks and otherwise—deaths of minorities fall in-line with those of whites. While in California, deaths of minorities relative to population remained elevated. ++ States w/high levels of BLM support that "follow The ScienceTM" are far more likely to have large numbers of children learning 100% virtually—which disproportionately impacts minorities. Once again politicization of science is causing people to betray their principals. ++ Beyond low-levels of in-person education and employment, masking is also linked to increased fear of illness, feelings of isolation, and spending nearly half as much time with others. ++ Nor is there any link between these supposedly "life-saving" behaviors, and decreased COVID deaths, either short- or long-term. A larger % of people spending time with others indoors is also NOT linked to an increase in deaths. ++ And yet the fear is almost insuperable, and leads those in blue states to wildly overestimate their risks associated with COVID. ++ The real risks are here--All data from the CDC, and un-obfuscated for your fear-destroying pleasure. ++ And yet politicians, and the Public Health and Medical establishments continue tell us that we must continue to mask—despite all this harm, and no benefit. Why? ++ The real question is why

Table 1: CDC Estimated Infection Fatality Rate (IFR) for COVID-19 Infections through 12/31/20 relative to 2017-2018 flu (worst recent flu year)

COVID Age group	Infections, thru 12/31/20	Symptomatic Illness, thru 12/31/20	Hospitzliations, thru 12/26/20	Deaths, thru 1/09/21	IFR All Infections	IFR Symptomatic Infections	Hospitalization Rate, symptomatic	Relative Risk
0-4 yrs	3,001,623	2,558,307	37,230	53	0.002%	0.002%	1.5%	1/2 as deadly as flu
5-17 yrs	14,550,829	12,403,731	100,007	201	0.001%	0.002%	0.8%	1/4 as deadly as flu
18-49 yrs	41,940,215	35,787,079	1,063,736	13,044	0.031%	0.036%	3.0%	2x as deadly as flu, slightly higher than risk of dying in childbirth
50-64 yrs	14,447,134	12,321,758	1,018,159	47,643	0.330%	0.387%	8.3%	6x more likely than dying in a car accident
65+ yrs	9,039,683	7,327,183	1,883,794	257,041	2.843%	3.508%	25.7%	4x more deadly than flu
All ages	82,979,484	70,398,058	4,102,926	317,929	0.383%	0.452%	5.8%	

Sources: COVID-19 Deaths Weekly Number of Deaths by Age (deaths through week of 1/15 included); COVID-19 Estimated Disease Burden; 2017-2018 Estimated Disease Burden, Influenza;

anyone continues to believe them. They lied about 6 feet, they lied about contagion from surfaces. The already told us they lied about masks—but the "confession" itself was the lie. ++ The CDC knows they are lying about masks—just as they have lied about the utility of all the other measures. The CDC itself published the following meta review in May 2020 based on all extant RCTs of masking and other NPIs. ++ Why then, do they keep saying that masks work? Ignoring every troves of data, and simply insisting "Masks work"? ++ The answer is 4-fold. First: Lockdowns had never been tried, and were not in the plan. The decision to implement them looks to have substantially increased deaths, creating an accidental genocide. ++ As well as a geriatricide, that is far worse simply than the nursing homes. ++ Dr. Fauci recommended masks about 3 weeks after they started, on 4/8. Presumably he was by then, well aware of the colossal fcuk-up, and using them to mask the enormity of this political and public health blunder. ++ The second reason is that the CDC has a really hard time giving up on droplet transmission—despite the massive amount of evidence to the contrary. This is because masks DO work to stop droplets, and aerosols over 5 microns. ++ Unfortunately, a large amount of work has been done on flu transmission, and as far back as 2008, it was shown that less than 0.1% of virus was in droplets over 5 microns, and 87% were below 1 micron. ++ So what about COVID? Is it also in such small aerosols? Well, in June of 2020, it was published that SARS-CoV-2 was found in aerosols in the 0.2-0.5 micron range—10-25x smaller than what can be captured by a mask. ++ Even the mechanistic studies show that with a 1% gap (far less than is present in an non-fit-tested N-95, and almost every single cloth or surgical mask, efficacy drops to 12% for an N-95. ++ If you're curious what a 1% gap looks like, it's here. This is equivalent to 1000 times larger than a 1 micron particle—the ones where they believe most virus is. That's 1/5th of a mile for you, and most gaps are 5-10x that size. ++ And yet the CDC persists with its mechanistic studies, cherry-picked data sets, and ignores empirical data, because it so desperately WANTS/needs masks to work. This is surely why they are so stern in their rebukes to states that lift mandates—they prove the lie. ++ The third and final reason is that for a long time, Public Health has gauged its success not by improving health, but in the uptake of its messages. This is akin to a company caring more about consumer recall of an ad, than the amount of product purchased. ++ And "masks work," has definitely sunk in. People believe it, just like they believed the sun revolved around the earth. It "makes sense," as long as you don't think about the physical characteristics of aerosols, or the data ++ Whether it works, is immaterial. "Masks work" WORKS FOR THE CDC, so they will stick with it, even if it puts others at-risk. And it does. ++ Think for instance of the "hairdresser" case study. What does it tell an at-risk person? It says: "It's fine to get your hair cut—so long as everyone wears a mask." In truth, this puts at-risk people at serious risk—I know, b/c my step-mom did just this—and got COVID. ++ More than any other reason, politicians and scientists-cum-politicians at the CDC (Dr. Walensky, literally reads from a script prepared by the White House) don't want the people who have mistaken them for gods to realize their fallibility, much less, their treachery. ++ They have caused grave harm, and they have caused the gravest harm to those who have the most trust in them—big-government, social justice-minded liberals. When those people realize they have been lied to so brazenly, there will be a cataclysm. ++ In many ways, this is like the clergy sex-abuse scandal. The amount of trust in clergy was so great, as to make them almost unassailable. But when Catholics realized how grossly their trust had been abused, it shook the Catholic church to its core. ++ Doctors and scientists are secular priests, and they have abused the trust we put in them. And like the priests before them, the harms caused by these abuses were visited upon the most vulnerable.

TEAM REALITY

Clifton Duncan: Good Looking Loser.
@cliftonaduncan

Those opposing Covid measures are NOT driven by ignorance & selfishness.

We actually understand the true risks of the virus, and want the most vulnerable to be taken care of.

We also want EVERYONE to retain the rights and freedoms they had prior to 2020.

11:12 PM · Mar 28, 2021 · Twitter for Android

The TRULY selfish, ignorant people are those who reject any data or opinions that clash with what they WANT to be true, and who expect the rest of the world to cater to their fears--so much so that even the suicides of CHILDREN mean NOTHING to them. And you call ME selfish??? ++ And just so you know, I say all this as one of the VERY few New Yorkers who was self-isolating & masking in Jan/Feb 2020; I was sanitizing my groceries and mail, for fuck's sake. I was 2 months ahead of the WHO in recognizing the pandemic. So MISS ME with your bullshit.

Eric
@The_OtherET

COVID cases in Midwest states that have not completely removed restrictions vs. states that have removed them since 1/1. If you squint enough, you can tell they are actually two separate lines.

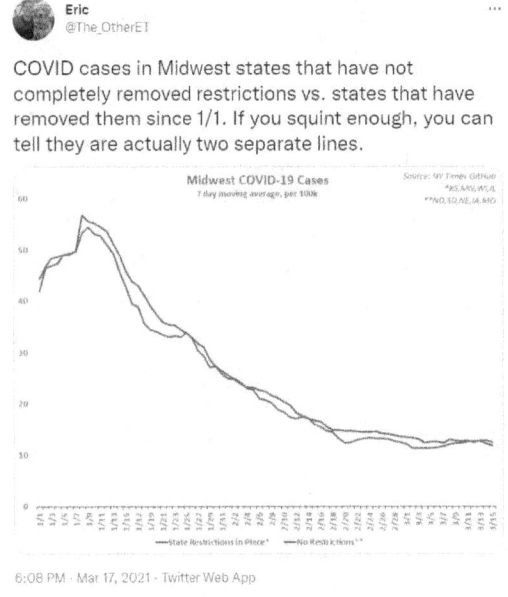

6:08 PM · Mar 17, 2021 · Twitter Web App

District Al
@districtal

Before you listen to anyone criticizing Senator Paul during his interaction with Fauci, remember that their last run-in consisted of Paul telling Fauci that schools were safe and Fauci disagreeing with him. Paul was proven right and Fauci was proven wrong.

6:22 PM · Mar 18, 2021 · Twitter Web App

Jennifer Sey
@JenniferSey

I have an "impending sense of doom" about kids entering their second year of virtual learning. Filling mental health facilities. Suffering from anxiety, depression and self harm.

4:32 PM · Mar 29, 2021 · Twitter Web App

Eric
@The_OtherET

Presented without comment

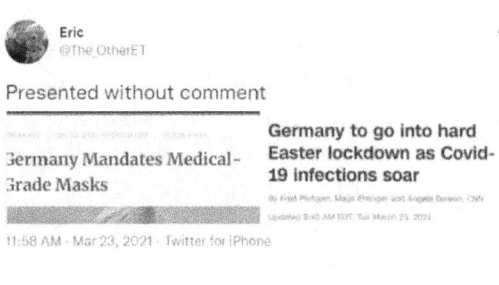

11:58 AM · Mar 23, 2021 · Twitter for iPhone

Wes Pegden
@WesPegden

While I'm not sure how harshly history will judge the timing and speed of 2021's spring relaxations in the grand scheme of things, I am confident that "absolutely reckless" will become the consensus description of decisions to keep schools closed to children for more than a year.

Gavin Newsom @GavinNewsom · Mar 2, 2021
Absolutely reckless. twitter.com/AP/status/1366...

4:27 PM · Mar 2, 2021 · Twitter Web App

Zac Bissonnette
@ZacBissonnette

There are Tweets circulating from very nice liberals explaining that they enjoy working from home and watching movies while low-wage workers bring them everything they need, with no interaction.

Let's just say I find these nice liberals less endearing than they find themselves!

4:26 PM · Mar 6, 2021 · Twitter for iPhone

APRIL 2021

Erich Hartmann
@erichhartmann

I wanted masks to work. I really did. We were in desperate need of a sense of 'control' over this situation, and our place in it. It's understandable to desire that.

What's not OK is a continued blind adherence to the COVID gods of *could* *may* and *possibly* in lieu of data.

8:45 AM · Apr 9, 2021 · Twitter Web App

TEAM REALITY

Bachman
@ElonBachman

Just flew from a deeply Blue US state to Dubai

The differences?

One is run by an unelected priesthood, disallows dining after sundown, has a curfew, and fines citizens for not covering their faces in public

The other is a pleasant constitutional monarchy on the Persian Gulf

4:58 AM · Apr 21, 2021 · Twitter Web App

Prof
@covidtweets

Told you so...

> **Prof** @covidtweets · Feb 24, 2021
> Looks like the Spring wave is starting in Europe, meaning we are likely 2-3 weeks away.
>
> This will be attributed to 1) variants and 2) people letting their guards down. Many will freak out.
>
> Don't...
>
> #openschoolsnow
> #endthechildabuse twitter.com/covidtweets/st...

2:07 PM · Apr 25, 2021 · Twitter for Android

Hold2
@Hold2LLC

CoV2 Neanderthal Update: 4/23/21

I'm seriously asking:

- Why should we believe the CDC and White House understand CoV2 transmission and risks?
- Why are masks still a thing after a year of showing no value?
- How have masks become MORE of a talisman?
- What data support masks?

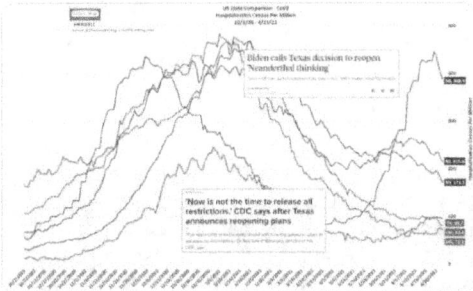

11:07 PM · Apr 23, 2021 · Twitter Web App

Ian Miller
@ianmSC

Remember when Phil Murphy said that he was "stunned" by Texas lifting their mask mandate, that he couldn't "fathom" it & that it took his breath away? Then Texas had way better numbers for 7+ weeks?

Super glad the media's been all over it, cause it's pretty embarrassing

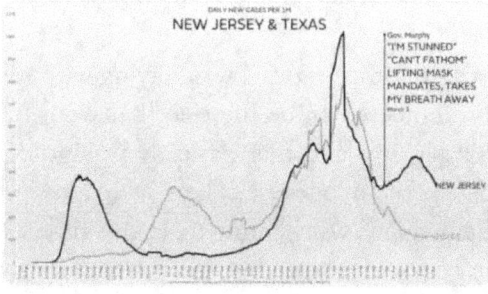

3:17 PM · Apr 27, 2021 · Twitter for Mac

Real Developments
@pdubdev

Public health officials have never presented evidence that masks or lockdowns work

No random controlled trials

No regressions

Just rhetoric and fear

11:44 PM · Apr 3, 2021 · Twitter Web App

Anish Koka
@anish_koka

The same ppl that spent a full year (and continue to do so) cultivating panic from what if COVID scenarios (sudden death, myocarditis, COVID brain, etc) based on random anecdotes are now shocked by others who raise the same concerns about vaccines.

8:30 AM · Apr 10, 2021 · Twitter for iPhone

Kyle Lamb
@kylamb8

"I would challenge whatever 24 year old intern that decided this was medical misinformation to a debate," Dr. Jay Bhattacharya says of the video being pulled from YouTube for saying he doesn't believe children should be wearing masks.

10:44 PM · Apr 8, 2021 · Twitter Web App

TEAM REALITY

District Al
@districtai

Full Disclosure (slightly personal and hopeful thread):

When all of this started about 13 months ago I was scared. I heard all of the stories coming out of China and then I heard and saw what was happening Italy. Thought no way it could come here like that.

1/x

9:33 AM · Apr 27, 2021 · Twitter Web App

In fact I was out on U street in DC the last Saturday night it was open. One of my undergrad degrees is history so I immediately thought this is the black plague of our time or spanish flu. I was hearing CFRs of 5-10%. The world went from light to dark almost overnight. ++ After hearing that I hated myself for endangering my family by even going out that night. I was staying with relatives close by as my place was getting finished. I volunteered to be the one that went out, that wore the mask, that wore gloves (yes i did! ha) ++ and that basically was going to be the one that got sick in the household instead of anyone else. There were kids in the house and i wanted to do my best to protect them. NYC was lighting up at this time. Then the Mid Atlantic started lighting up. Again fear consumed me. ++ I engrossed myself in reading about virology, CFRs, IFRs, etc. I started reading about the methodology of these doomsday models (ive done predictive modeling for 12 years so I was more educated than most on them) then the end of April rolled around.... ++ Yes Covid is/was real but the models didnt make any sense. Marrying the textbooks I bought with the data science I knew, something just didnt add up. Slowly I started looking at the data myself. I started doing my own research which is what i always told my.... ++ grad students to do when i was a professor. I started pulling my own data, following every account I could, reading everything i could to make my own decision. The data was not telling me what the media was telling me. The data had patterns. The media had fear. ++ At that point I personally was past fear. I started seeing the lack of efficacy for masking, the age segmentation of covid, ineffectiveness of lockdowns as Covid just traveled a natural path around the country. ++ The data does not lie and we have spent a year ignoring it. I get that people do not want to believe the data that tells them it isnt as bad as you thought. It took me a month or so to believe it and i do this for a living plus im getting my MPH in epi. ++ If it isnt as bad as you thought then your actions didnt mean what you thought they meant. That is a massive blow to some people and i get that but it is time to let go. No one is mad at you but it is time to look at the data objectively and see we are winning.... ++ We are getting past this. Do not be stuck in April 2020, come into April 2021 because May 2021 will be triumphant. The water is warm and you are welcome. It is time. Shake a hand, smile at someone, give someone a hug. Time to live.

Kyle Lamb
@kylamb8

Just announced at a presser: Florida is filing a lawsuit against the federal government and CDC in order to reopen the cruise industry

11:59 AM · Apr 8, 2021 · Twitter Web App

Prof Francois Balloux ✓
@BallouxFrancois

"Arguing that you don't care about the downsides of vaccine passports because you have been, or are being, vaccinated is no different than saying you don't care about health discrimination because you are healthy and financially well off."

4:06 PM · Apr 3, 2021 · Twitter Web App

TEAM REALITY

Josh Stevenson
@ifihadastick

There are more students in Virtual only school in California than in all other states combined.

END

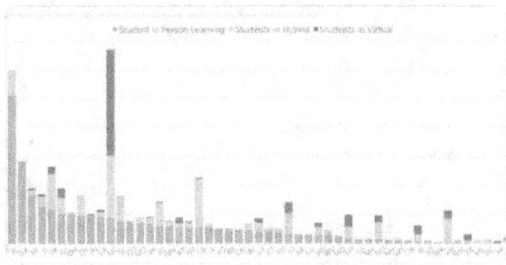

9:44 PM · Apr 19, 2021 · Twitter Web App

Gummi Bear
@gummibear737

Imagine if health "experts" had used the Covid-19 crisis to scare people into losing weight and excercising (and taking vitamins)

Would have been the most positive health initiative in history

Instead they went for sedentary lockdowns

Lives saved propably negative

7:28 PM · Apr 17, 2021 · Twitter for iPad

John Ziegler ●
@Zigmanfreud

The weirdest thing about mask mandates becoming a tenant of the woke religion is they actually violate several of their stated beliefs:

-My body, my choice
-No to governmental control/yes to civil liberty
-Abusing children is always unthinkable
-All environmental harm is wrong

9:34 PM · Apr 22, 2021 · Twitter for iPhone

John Ziegler ●
@Zigmanfreud

We've reached a key moment in the quest for "normal." Its clear that those who don't want normal are bizarrely downplaying vaccines, while also blackmailing people into getting them.

I fear, for mostly structural reasons, data will NEVER get good enough for freedom to fully win.

10:26 PM · Apr 13, 2021 · Twitter for iPhone

Prof
@covidtweets

When I was writing this, @CDCDirector was warning us of an impending doom.

When you ignore seasonality, you will be wrong.

Every. Single. Time.

Kudos to @districtai and @Hold2LLC too; this "model" was a product of our combined brains.

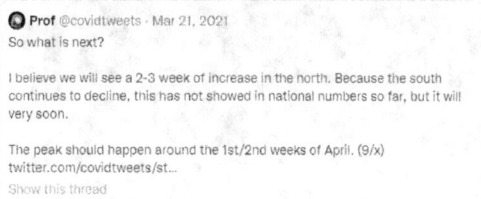

> **Prof** @covidtweets · Mar 21, 2021
> So what is next?
>
> I believe we will see a 2-3 week of increase in the north. Because the south continues to decline, this has not showed in national numbers so far, but it will very soon.
>
> The peak should happen around the 1st/2nd weeks of April. (9/x)
> twitter.com/covidtweets/st...
> Show this thread

7:44 PM · Apr 22, 2021 · Twitter for Android

Tracy Hoeg, MD, PhD
@TracyBethHoeg

Stunning:
1. This high school junior ran an 800m in 2:08 - in a mask
2. She was *forced to run this outdoors in a mask*

Not stunning:
That she collapsed

What are we doing to these kids?
This does 0 for pandemic control
WHO says no masks during exercise

news.yahoo.com/high-school-co...

7:26 PM · Apr 27, 2021 · Twitter Web App

Justin Hart
@justin_hart

If it wasn't for the government I wouldn't know what to do with myself

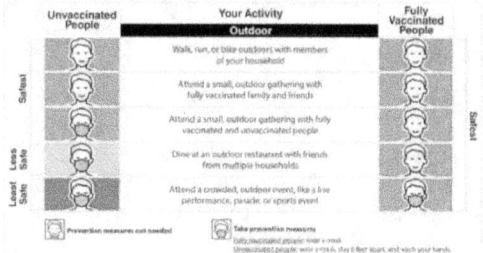

12:35 PM · Apr 27, 2021 · Twitter Web App

Angry Cardiologist
@AngryCardio

I sure hope that students attending colleges & universities that are mandating COVID vaccination for school re-entry don't experience any adverse events.

11:19 PM · Apr 26, 2021 · Twitter for iPhone

TEAM REALITY

Pajamas It Is
@HeckofaLiberal

Every person in this picture has been fully vaccinated. Their messaging is very poor. We need better public health experts. Two administrations now doing opposite things but both very wrong. Wake up America.

5:47 PM · Apr 11, 2021 · Twitter for iPhone

Prof
@covidtweets

Here is a Venn diagram showing the states that recently lifted their mask mandates and those currently experiencing COVID surges.

11:38 PM · Apr 14, 2021 · Twitter Web App

Emma Woodhouse
@EWoodhouse7

Are you vaccinated?

Fine.

Then ACT LIKE IT.

7:52 PM · Apr 23, 2021 · Twitter for iPhone

Wes Pegden
@WesPegden

TFW you read the Twitter threads where epidemiologists come out and say

"yah, wearing masks outside has been pointless and/or harmful, maybe the government should stop making people do it"

and the people respond with

"No! The government must make me do it! Why not, after all!"

9:02 PM · Apr 19, 2021 · Twitter Web App

Craig
@TheLawyerCraig

Replying to @districtal

I will post this technical chart in the thread tonight, but we are currently somewhere around "Yay" right now, though some states will start moving toward "Oh no" as early as today. We should be at peak "Oh no" in raw metrics this ~Thur and in 7DA metrics the following ~Wed/Thur.

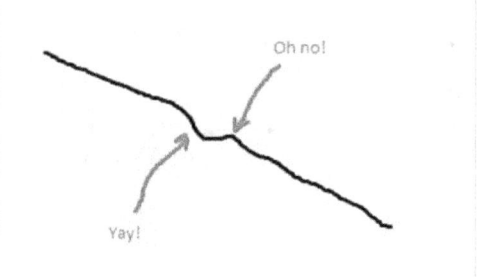

4:49 PM · Apr 5, 2021 · Twitter Web App

Wes Pegden
@WesPegden

I'm still kind of blown away that instead of

"the vaccines are amazing, will cut transmission, turn even variant COVID into a cold; so get them and get back to life!"

we went with

"we're not sure how good they are, and we're going to make you get them if you want to do stuff."

6:26 PM · Apr 5, 2021 · Twitter Web App

Federico Andres Lois
@federicolois

Replying to @federicolois @MrImplausible and @MLevitt_NP2013

The day you stop asking questions, is the day you become the most anti-vax of all. The pro-vaxxer (I call them stupid-vaxxers or vaxidiots) that some day will allow some big incident to happen that would hamper vaccination campaigns and fuel the antivax movement for years to come

12:53 PM · Apr 28, 2021 · Twitter Web App

ClownBasket
@ClownBasket

Gen Zers born between 1998-2006 are the biggest casualties of Covid. HS Freshman thru College Seniors.

Many have not seen a classroom in over a year. Masked sports (if any). Social restrictions.

Adults see them a transmitters of disease; not as the future to be invested in.

6:02 PM · Apr 6, 2021 · Twitter for iPhone

TEAM REALITY

My relevant experience is as a district education director responsible for integration of immunocompromised, profoundly disabled, undocumented, Autistic, and behaviorally challenged students under full ADA, IDEA, and OSHA compliance, with a background in hazardous environs 2/ ++ PPE applications, which includes which respirators work when and why. My experience grants me the capacity, for instance, to understand specifically how none of you sitting before us today are protected against Covid size particulates, nor are you protecting others - but you 3/ ++ know this already, don't you? It is critical information to consider that there are ZERO efficacy standards for child size masks. You are requiring untested, unregulated apparatuses which restrict breathing, cause increased carbon dioxide, and are a petri dish of secondary 4/ ++ pathogen directly in front of oral/nasal mucosa. You tell me if you think this sounds like something that would pass an ethics review board. We're going to take a bunch of kids and put them in this apparatus, ok? They'll be in it for 8 hours per day, up to 12 if they are 5/ ++ in extended day programs. We will not ask for medical clearance or medical consent, but this apparatus will cause deoxygenation and hypercapnia, and children being the sanitary creatures that they are, will drop them, wear them into restrooms where they pick up fecal matter 6/ ++ and other pathogen from toilet plumes. I hope each of you who have worn a mask in a public restroom today recoil in horror at the realization of what you're breathing - tasting, even - all day long. Kids wear them all day. To be exempt, they must fight tooth and nail and face 7/ ++ rampant rejection. Would this pass an ethics review in your opinion? Of course it wouldn't, but here we are, and this is what you are doing to children in this school system every day, even our truly immunocompromised, in unregulated, unsafe garbage which exacerbates the 8/ ++ spread of airborne pathogen. But we're all in this together, right? Cloth and surgical masks are expressly non-mitigating for airborne pathogen. Covid has a minimum particle size of .06 microns, which if part of a larger cluster still easily falls within the radically behaving 9/ ++ particulate range. Add in plosive force, which is the varying outward respiratory pressure, and you have much like water through a garden hose on the mister setting - droplet into aerosol - where the tighter the fit, the greater the pressure of fine particulates in an outward 10/ ++ plume, effectively taking what falls in a predictable 6 foot arc and sending it an 18-20 foot trajectory, remaining aloft for hours in enclosed spaces. 90% of exhaled emitted particulates fall within the radically behaving particulate range - what you have is truly an airborne 11/ ++ pathogen. You say you'll listen to CDC guidance, but cherry-pick which, ignoring how 71% of new cases report always masking, 14% most of the time. **Buzzer** What else I would have said: There is a developmental requirement of seeing lip, tongue, and tooth placement for 12/ ++ linguistic onset and utterance progress, as well as the ability to emulate caregiver social cues. Missed milestones can take years to remediate, if remediation is even achievable. What ever happened to IDEA Least Restrictive Educational Environment requirements, 13/ ++ segments 300.114-300.120? We cannot cherry-pick which federal laws and regulations to follow while ignoring the rest. By following these practices, there is the creation of a heightened discriminatory atmosphere for employees without accommodation required under ADA federal 14/ ++ workplace integration access law, wherein mere presence does not constitute direct threat, even if sick with a contagious pathogen - which is why you had bo issue getting flu meds inside a Target or Publix PreCovid. These practices are OSHA noncompliant for, as workplace 15/ ++ respirator use for known pathogen requires medical clearance, medical consent, and a fit test of new PPE correctly matched to the pathogen provided by the employer. None of the people you are mandating masks for have been allotted the medical consent and clearance required. 16/ ++ Furthermore, never before were pregnant employees mandated to wear deoxygenating apparatuses that also cause dramatic spikes in CO_2 levels (over 10,000 parts per million within the first 90 seconds of wear, far over allowable exposure concentrations) for the duration of their 17/ ++ pregnancies. End mandated masks in district schools immediately. 18/

Matt, Pre-School Diploma
@statomattic

My realization after the J&J decision has marinated 12 hrs. -

All my life, whether I liked those in charge or not, I trusted the US gov't to be at least broadly truthful.

But after the past year, I don't. At all. They have lied/manipulated so much through the pandemic... (1/3)

8:07 PM · Apr 13, 2021 · Twitter for Android

...that anything they tell me, I instantly doubt. I now view our government and media as propaganda outfits with 0 respect for the public. Anything they say - ANYTHING - I feel the need to seek out primary sources and do my best to guess whether it is true, or what IS the truth. ++ I am in no way a conspiracy theorist. I am a person who desires facts, whose trust has been betrayed enough so that I no longer trust. And to feel like I can no longer rely on what should be trusted sources is wearying. I can't be the only one. This is scary, sad, and very bad.

Victoria Fox
@drvictoriafox

I can't believe I have to say this but Vaccines are the end point for the COVID pandemic. After every adult who wants to be vaccinated has been vaccinated that's it - the end - whatever the outcome we live with it and go back to normal.

7:25 PM · Apr 14, 2021 · Twitter for iPhone

AJ Kay
@AJKayWriter

The sheer desperation in these coercive vaccination policies and marketing pushes should make any rational person wonder why the powers-that-be want this so bad.

The harder they push, the more holdouts will resist.

Trust has been decimated...and legitimately so.

1:13 AM · Apr 28, 2021 · Twitter for iPhone

TEAM REALITY

ClownBasket 🤡
@ClownBasket

1/13

UPSIDE: A THREAD

There has been some UPSIDE to the tragic hijacking of science and the tyrannies affecting civil liberties over the last year.

(disclaimer: any UPSIDE has not been "worth it" (yet). Nor is this a cost-benefit analysis.)

Let's take a gander...

9:38 PM · Apr 22, 2021 · Twitter Web App

Public Education. Pretty much everyone hates the Teachers Union now. I certainly was not aware of their stranglehold on public schools. Unions will pay a price for overplaying their Covid hand. Competition is good for innovation & accountability. ++ Local elections. I had no idea our local School Board could keep our kids out of school and then blame & shame them for Covid while obsessing over contact tracing & quarantines. Local school boards are the most important elections in the Republic. ++ Gov Ron DeSantis. A "presidential" Trump? I must admit I didn't know anything about him 12 months ago. Now I feel like he should run for President. He has the midas touch. ++ Libertarians. I kinda always thought about libertarians as a quirky group of people who didn't want to wear seat belts. Now they seem to be the least authoritarian party. 4 years ago I voted for Hillary. Am I a libertarian now? ++ Gap Year(s). Academia is a sham. Where were all the brave universities and their scientists speaking out against bad science and defending liberties? Helpful tip: Universities are a waste of money. ++ Scambusting. God forbid there should be another attempt at a coordinated scam. Team Reality has used social media to coalesce around data and evidence. Credentialism is dead. ++ 1776. Critical Race Theory (CRT) is a sneaky beast. I had NO IDEA how bad it was. In fact, I learned a lot about CRT over the last 12 months. This racist propaganda is destroying education. Now we know. ++ Randomized Controlled Trials. Those outside of the research fields have gotten a good education on what quality research looks like. Oh and now we actually read the studies. ++ Anti-vaxers. I am not anti-vax; I am just pro-test. Hence, I will wait on these gene therapies. BUT I better understand the anti-vaxers now. Empathy. ++ Engagement of Gen X Parents. They pushed us too far when they messed with our kids' mental health. We might have a bad reputation, but we are pragmatic and come prepared to the school board meetings. Evidence, bruh. ++ Competition for Big Tech. The censorship is out of control. The solution is probably not Section 230, but instead the rise of other platforms that support the values of the Republic. I am ready to support Patriotism and the 1st Amendment. ++ The Activation of iGen (Gen Z). Abused by every generation that proceeded them, locked out of school, activities, & relationships, and witnessed adult double standards & hypocrisies in real time. They will save the world, I am sure.

Stefan Baral ✓
@sdbaral

Just so I'm clear, are we planning on vaccinating young children to protect vaccinated adults?

9:40 PM · Apr 15, 2021 · Twitter Web App

TEAM REALITY

 **Emma Woodhouse**
@EWoodhouse7

Truly incredible. Illinois % positivity for PCR tests peaked on the SAME DAY last day & this spring: April 12th app.powerbi.com/view?r=eyJrljo...

Seasonal
Endemic
Survivable

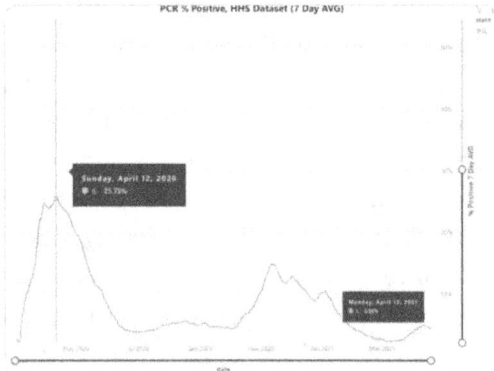

7:36 PM · Apr 22, 2021 · Twitter Web App

 **Bethany S. Mandel**
@bethanyshondark

Our local library did this cute thing where they put children's books in the window and said "call us and we'll bring them out for you." It's great but you know what's better? OPENING THE FORKING LIBRARY.

5:09 PM · Apr 27, 2021 · Twitter for iPhone

 **Martin Kulldorff**
@MartinKulldorff

Boston is another example where lockdowns have protected the educated laptop class at the expense of the working and middle class.
Analysis/graph by @Emily_Burns_V.
rationalground.com/lockdowns-syst...

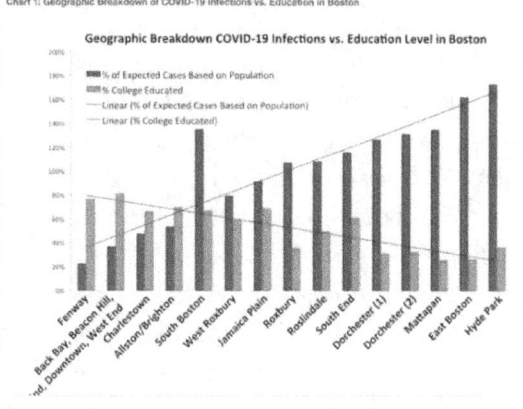

11:36 AM · Apr 15, 2021 · Twitter Web App

Link for the article quoted ☞:

https://rationalground.com/lockdowns-systemic-racism-in-action/

 **Eric**
@The_OtherET

United States map marked with the 10 states with the highest COVID hospitalization rates as of 4/10

Now, I may not have 7 acronyms after my name, but I think I'm starting to pick up a pattern here

10:41 AM · Apr 11, 2021 · Twitter Web App

 **elizabeth bennett**
@ebennett74

As a child advocate, the forced, invasive nasal swabbing of well school children to appease irrational adult fears makes my blood boil. It should make yours boil too.

10:24 AM · Apr 19, 2021 · Twitter for iPhone

 **Eric**
@The_OtherET

It's not selfish to want to wait for full authorization of the COVID vaccines, or to wait for more long-term safety data on them before taking one

It is selfish, however, to coerce others into receiving one for your "safety" when you are perfectly capable of getting one yourself

11:08 AM · Apr 7, 2021 · Twitter for iPhone

TEAM REALITY

Emily Burns 🟢 DMs welcome #TeamReality
@Emily_Burns_V

1/
Masks don't work to stop COVID, though they work wonders keeping kids out of school & unemployment high

But WHY don't they work? Let's see

Rather than thinking of respiratory aerosols like tiny spit globules, a better metaphor is cigarette smoke

1:11 PM · Apr 9, 2021 · Twitter Web App

elizabeth bennett
@ebennett74

"Dear primary care doctor, please be advised that your patient has been seen at our hospital for a behavioral health admission."

Sure getting tired of these. The kids are not all right.

#OpenAllSchoolsNow #5FullDays

7:34 PM · Apr 23, 2021 · Twitter for iPhone

Eric
@The_OtherET

When you plot the current COVID stringency rating of the 50 US states against their current hospitalization rates, there is next to no correlation and what little correlation there is goes in the opposite direction it's supposed to

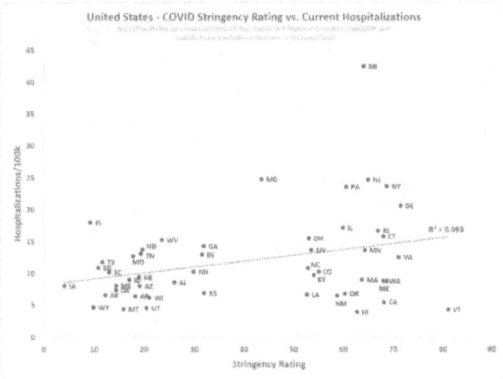

4:12 PM · Apr 19, 2021 · Twitter Web App

Zac Bissonnette @ZacBissonnette · Apr 27, 2021
This is just hilarious.

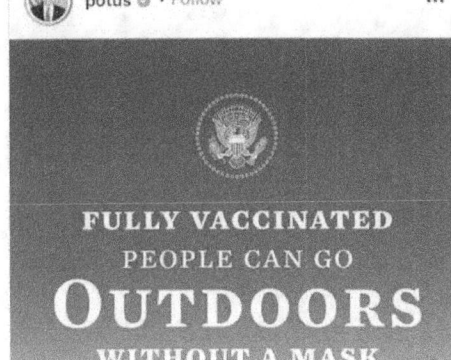

Jennifer Sey ✓
@JenniferSey

Masking outside is performative.

Masking small children is performative.

Masking small children outside is lunacy.

I won't submit.

8:02 PM · Apr 25, 2021 · Twitter for iPhone

Bethany S. Mandel ✓
@bethanyshondark

I think we've learned in the last year how incapable everyone, from the FDA/CDC on down to the regular population, can conduct a rational risk analysis.

9:10 AM · Apr 13, 2021 · Twitter for iPhone

MAY 2021

Emma Woodhouse
@EWoodhouse7

I'm starting to think we don't all share a common understanding of what the words "guidelines" and "recommendations" mean.

9:15 PM · May 27, 2021 · Twitter for iPhone

NYC Angry Mom
@angrybklynmom

I have to make some confessions.

Last year, I was absolutely insane and I was one of those people who yelled at people outside for not wearing a mask.

It wasn't actually the virus that I was afraid of, but the lockdowns.

7:35 PM · May 18, 2021 · Twitter for Android

I was so deeply psychologically damaged from the NYC lockdown in the spring that I lived in perpetual terror of it happening again. Cuomo spent the entire summer & fall threatening and browbeating us for our "bad behavior" and I deeply internalized this message. ++ I remember screaming at my neighbors and telling them that if they didn't put on a mask, they were going to face my wrath if my kids' school was closed. At one point, he put us into an arbitrary microcluster and we escaped closed schools by only a few blocks. ++ I spent the entire school year watching every stupid metric and desperately trying to get three steps ahead of Cuomo. I was very lucky that my children's (Catholic) school was never closed. One child had 2 quarantines (and one was over a break). We got incedibly lucky. ++ To be clear, I never supported the lockdowns. I thought they were an abomination. Pure psychological torture. Masks felt like the compromise to keep us out of lockdown, & I got angry when everyone wouldn't "comply", to keep us out of lockdown. ++ If you go back and watch Cuomo's briefings, the threat of closures (dependent on our behavior!) were a daily phenomenon. And when they happened, they were metric-free, indefinite in nature, arbitrary and capricious. Completely at the whim of one man, and one man only. ++ This thread took off, so I want to clarify a few things. The reason I wasn't as personally concerned about the actual virus as I was about the lockdowns is bcs: -I felt confident that I knew how to protect my family -Data already showed the severe age stratification ++ But lockdowns were something completely out of our control. We opened the door to something new & nefarious - (inept) govts having 100% power over its citizens' doings, with fear as a powerful motivator for compliance. My fear was that we might never fully claw back that power ++ I had no trust in our gov't's competence, nor in their ability to put the well being of citizens ahead of their own aspirations & desire for power. I had no trust that they were consulting the right experts, or were effectively leveraging the data to make informed decisions. ++ The psychological terror was in conflict with the logical part of my brain - this part understood that any govt in NY's position during Spring 2020 would have no choice but to take serious action in the face of a threat we knew very little about. ++ But when we reached the tail end of that first terrifying wave, it seemed obvious that we should have been taking a step back and designing a long term plan to live with the virus until a vaccine could be produced, which addressed the full spectrum of human need. ++ It shouldn't need to be explicitly stated that it's not sustainable (or humane) to demand people sacrifice: -Time with loved ones -Education -Livelihood -Socialization For months on end, or even YEARS. Yet, this was/is unapologetically the expectation in many parts of the US. ++ To this day, this is still an expectation in some areas, even after having a year+ worth of data from restricted vs open states proving that these profound sacrifices are not correlated to better viral outcomes. And even after the population having broad access to vaccines.

TEAM REALITY

Abir Ballan
@abirballan

Children should not be vaccinated for the wrong reasons.
They have almost zero risk from C19 and so they get zero benefit from the vaccine.
Do not vaccinate your child for C19:
1) just to conform and be pleasant
1/n

> Abir Ballan @abirballan · May 25, 2021
> «The attention should be on the institutional review boards of the universities who are approving injecting literally experimental drugs. The vaccines are unde emergency use authorization.» Dr Scott Atlas 3/n
> Show this thread

5:15 AM · May 29, 2021 · Twitter Web App

2) to be able to send them back to school (children have the right to pursue education regardless of their vaccination status
2) to protect the vulnerable. The vulnerable have several options to protect themselves.
3) to be able to travel easily or get your freedom back ++ Vaccination should only happen for medical reasons when the benefits outweigh the risks of the intervention. Protect your child from Big Pharma sale targets.

Vinay Prasad, MD MPH
@VPrasadMDMPH

Remember when the CDC said schools need 6 ft of distance and need to be linked to community spread, and we were like that has been contradicted by evidence, and then the emails were released between them and the teachers unions...

Yeah. That happened too

10:10 PM · May 4, 2021 · Twitter Web App

Ian Miller
@IanmSC

I think my favorite part of 2021 so far is how increasingly desperate the mask religion's arguments are getting as more and more evidence accumulates that they've been completely useless

The refusals to accept reality have gotten more and more unhinged and frantic

6:14 PM · May 27, 2021 · Twitter for iPhone

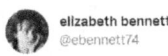
elizabeth bennett
@ebennett74

A great thing for public health officials to do would be to stop misrepresenting, exaggerating or straight-up lying about the risk of COVID to kids in order to try to scare parents into wanting to vaccinate them. Thanks for coming to my TED talk.

5:46 PM · May 26, 2021 · Twitter for iPhone

Angry Cardiologist
@AngryCardio

Lottery tickets to induce vaccine administration is what you get when you suck every last drop of human dignity out of governance.

10:45 PM · May 12, 2021 · Twitter for iPhone

Tracy Hoeg, MD, PhD
@TracyBethHoeg

Am I the only one who feels it is not appropriate to require kids to be vaccinated for a disease that poses very little risk to them with a vaccine that only has EUA to be able to take off their masks "outdoors"? 🙄 (we have data masking kids outside does 0)

9:31 PM · May 5, 2021 · Twitter Web App

Phil Kerpen
@kerpen

"But how will I know who's vaccinated???" people 100% missed the point. If *you* are vaxed you don't have to sweat it. You're good. Leave everyone else alone.

6:23 PM · May 13, 2021 · Twitter Web App

Justin Hart
@justin_hart

Never forget that the strongest pro-mask evidence the CDC could come up with was that they "slow the rate" of growth of cases for counties that required mask mandates for restaurants but only during the third and fourth week after the mandate or something and then less than 2%

11:17 AM · May 25, 2021 · Twitter for iPhone

Vanessa
@vlal42

So kids were forced by adults to isolate, mask up, endure e school to protect grandma. And now that grandma is vaccinated- they still have to distance and mask up? Where the hell are we doing here? Aren't we supposed to let the lowest risk return to normalcy?

10:09 AM · May 28, 2021 · Twitter for iPhone

TEAM REALITY

Jennifer Cabrera @jhaskinscabrera

This CDC guidance to stop wearing masks if vaxed exposes the fact that many organizations only followed CDC guidance because it reinforced what they wanted to do, anyway.

Many of those same organizations are NOT following CDC guidance now. 🤦

10:33 AM · May 14, 2021 · Twitter for iPhone

Real Developments @pdubdev

In the future it will be shown that lockdowns made Covid worse .

Caused elderly to be more exposed. Prevented young from developing blocking immunity. Shut down life saving hospital treatment, weakening and making many more susceptible.

Panic always make things worse

1:55 AM · May 16, 2021 · Twitter Web App

District Al @districtal

So now anti maskers are pro science and pro maskers are anti science right?

3:09 PM · May 13, 2021 · Twitter Web App

Tracy Hoeg, MD, PhD @TracyBethHoeg

Finally proof: The @CDCDirector had been "preparing to write that schools could provide in-person instruction regardless of community spread" when teachers unions intervened. Feb is when everyone at the CDC started to ignore the findings of our WI study 😬

7:38 PM · May 1, 2021 · Twitter Web App

John Ziegler @Zigmanfreud

After flying for 1st time since mask insanity took over, its very obvious airlines 100% embraced magic masks as a way to salvage their business. Because social distancing is impossible, too many people need a fake security blanket to be able to fly. I dont see how this ends soon.

10:00 AM · May 22, 2021 · Twitter for iPhone

Michael Tracey @mtracey

Now that "lab leak" has more-or-less gained complete mainstream acceptance as a plausible COVID origin theory, please look this post with examples of how arrogant journalists deceptively colluded throughout 2020 to dismiss the theory as "debunked" on the basis of jack-shit

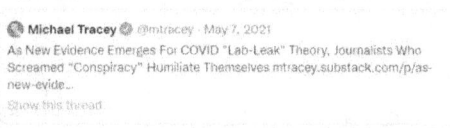

10:22 AM · May 18, 2021 · Twitter Web App

Don Wolt @tlowrion

The CDC's guidance can often seem incoherent, so here I offer a short list of environments where research shows masks have not been found to be effective in protecting the wearer or those around them from transmission:
- outdoors
- indoors

1:58 PM · May 14, 2021 · Twitter Web App

Stefan Baral @sdbaral

Just so I am clear, are we masking little kids as source control to protect vaccinated adults?

10:00 AM · May 28, 2021 · Twitter Web App

Ian Miller @ianmSC

It's hard to believe that we're still arguing about the effectiveness of masks & interventions after Florida essentially removed all restrictions, faced huge expert criticism and then had better results than lockdown states for 5+ months

Results are in guys, The Science™ failed

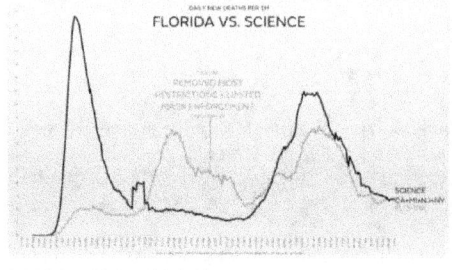

4:01 PM · May 19, 2021 · Twitter for Mac

---

elizabeth bennett @ebennett74 · May 24, 2021

Yes. I am so sick of hearing "what's the big deal, my kid doesn't mind." How fortunate for you if your kid is not special-needs, hearing impaired, still learning language, etc. Perhaps spare a thought for all the other children.

TEAM REALITY

Mark Changizi
@MarkChangizi

Your life since March 2020 in a nutshell.

1:05 PM · May 25, 2021 · Twitter Web App

AJ Kay
@AJKayWriter

Right now, the scorecard for predicting the last 14 months looks something like:

"Conspiracy theorists" - 10
Modelers - 0

Who should we really be listening to?

8:20 AM · May 24, 2021 · Twitter for iPhone

Stinson Norwood
@snorman1776

Which one forced masks onto children and prompted calls for mandated vaccinations?

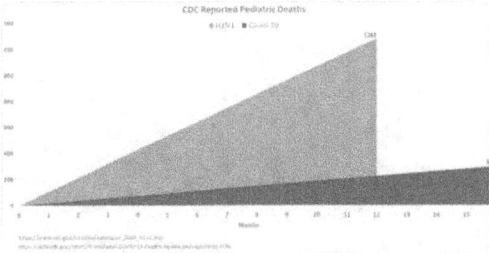

1:30 PM · May 27, 2021 · Twitter for iPhone

Phil Holloway™
@PhilHollowayEsq

When child abuse becomes normalized 😢

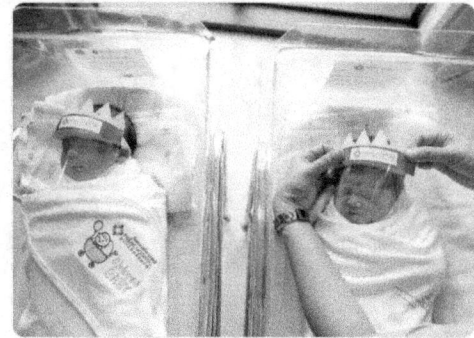

11:01 AM · May 10, 2021 · Twitter for iPhone

Phil Holloway™
@PhilHollowayEsq

Good morning ☀ as you get ready for your day, take a moment to think about all the little kids who are forced to suffer 7 hours + with their breathing restricted as they sit in schools across America:
twitter.com/MelanieMusey/s...

7:05 AM · May 7, 2021 · Twitter for iPhone

Vinay Prasad, MD MPH
@VPrasadMDMPH

When I see CDC twisting into pretzels with conflicting messaging, I think:

That is what happens when you distort science to try to get people to do what you wish they did

We are not good at behavior manipulation

Best to tell truth & not try it
medpagetoday.com/blogs/vinay-pr...

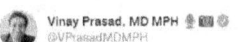

As I stated at the outset, I have profound respect for Fauci for his career of service, and like many, I am a fan of his clear public speaking. Yet, these two events force us to ask whether fact manipulation is acceptable.

I believe it cannot be. The public will not trust us, and should not trust us. People will put our statements through a reverse translator to try to deduce what we truly think, and it gives an unjustified power to scientists that belongs in the hands of people.

I can't control what others will do, but I can control myself. If I tell you what I think, I can't promise I am correct (in fact, like all mortals, I am occasionally wrong), but I can promise you that is, in fact, what I think.

Vinay Prasad, MD, MPH, is a hematologist-oncologist and associate professor of

11:23 PM · May 5, 2021 · Twitter Web App

Eli Klein
@TheEliKlein

Now The NY Times says "Immunity to the coronavirus lasts at least a year, possibly a lifetime"

"Experts" were critical of me last year when I told them that people who already had Covid shouldn't be subject to restrictions or masking

I have those receipts & they're not pretty

6:02 PM · May 26, 2021 from Manhattan, NY · Twitter for iPhone

TEAM REALITY

Show Me The Data
@txsaith2o

My kid got sent to the principal due to her club

She was told "You have to talk to the school board"

She talked to the school board

The next day @henrymcmaster circumvented mask mandates in schools

She's walking into school tomorrow like a boss

9:06 PM · May 11, 2021 · Twitter Web App

Matt, Pre-School Diploma
@statomattic

Get vaccinated and get free donuts!
Get vaccinated and win a college scholarship!
Get vaccinated and win a million dollars!
If you call RIGHT NOW, we'll give you an extra non-stick pan, free!

Hey, politicians: Want more vaccinations? Start acting like they work. Lose the masks.

8:00 AM · May 13, 2021 · Twitter for Android

David M
@ComradeDoom1

The wet market theory was actually the racist one. It was an attempt to scapegoat the dietary habits of ordinary people and provide an alibi for the depravity of elites.

10:13 AM · May 27, 2021 · Twitter Web App

ClownBasket
@ClownBasket

BREAKING: CDC adjusts the Herd Immunity Threshold to 120%

11:57 PM · May 5, 2021 · Twitter for iPhone

Vinay Prasad, MD MPH
@VPrasadMDMPH

It is almost that time of year where everyone who was mad at Emily Oster for saying you can go on summer vacation with unvaccinated kids starts posting pictures of their summer vacation… With kids

10:11 PM · May 22, 2021 · Twitter Web App

AJ Kay
@AJKayWriter

The whole "flatten the curve" thing really got away from us, didn't it?

6:12 PM · May 28, 2021 · Twitter Web App

elizabeth bennett
@ebennett74

Seeing a 14-mo old in clinic who hasn't had a checkup since she was 2 mo old (!) because the parents were terrified to take her out of the home. She is OVERDUE for 6 vaccines—all of which are for pathogens much more dangerous to her than COVID. Fear-mongering is a dangerous game.

12:44 PM · May 20, 2021 · Twitter for iPhone

ClownBasket
@ClownBasket

I will spell it out: SE Asia had SARS-CoV-2 in circulation as far back as 2018.

This explains the low case/death numbers out of CCP/SE Asia in 2020 (acquired immunity over 2 prior years).

So why the hysteria Dec '19? And why did the West buy in?

12:53 PM · May 7, 2021 · Twitter for iPhone

Kyle Lamb
@kylamb8

Update.

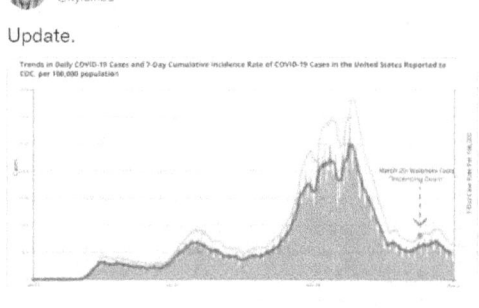

11:08 AM · May 5, 2021 · Twitter Web App

TEAM REALITY

ClownBasket 🤡
@ClownBasket

ThrowBack Friday to May 7 2020 - the day I came out of the Covid closet.

I wouldn't be active on Twitter for another 3 months and NextDoor was lame (censored 2x for opposing masks). So I crafted this email and sent it to every politician and media source I could find.

> If you had told me 8 weeks ago that today, in America, drones would be used to tell citizens to disperse, that you would not be allowed to leave your house but for certain permissible activities, that churches, synagogues and mosques would be closed but liquor stores would remain open, that you could not gather in groups to protest, that you could get arrested for going to the beach or playing in the park with your kid, that neighbors would be encouraged to report to authorities enforcement of "laws" NOT made by our legislative branch, that women could go to a clinic to get an abortion but not a mammogram or biopsy, that only certain people would be allowed to work, that you would be allowed to go to certain stores but restricted from purchasing certain things there, that small business owners would not be allowed to open to customers but still would be required to pay rent, utilities and taxes, that youtube and facebook would shut down anything they considered "misinformation," that people would throw around the words "science" and "data" for the purpose of shame, humiliation and control, that we would focus all our attention on one metric (Covid) at the expense of all others (child abuse, spousal abuse, influenza, suicide, mental health, cancer, etc.), that a movement toward autocracy would come not from the a demagogue president but from state governors and city mayors, that the Bill of Rights and the entire system of government that we leveraged for almost 250 years would be put on hold, and that all of these realities, characteristic of countries and regimes that are not democracies and do not value liberty or the individual, would drift into existence with little to no fanfare, press, or protest, I would have told you that you were crazy.

10:20 AM · May 7, 2021 · Twitter Web App

David M
@ComradeDoom1

a horrific icon of our society in 2021: young children muzzled to assuage the hysterical fears of geriatric adults.

11:35 PM · May 14, 2021 · Twitter Web App

Angry Cardiologist
@AngryCardio

Hey @Twitter, is it time to reconsider your ruling on my tweet?

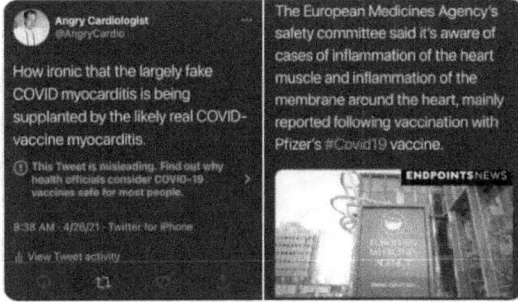

12:31 AM · May 8, 2021 · Twitter for iPhone

Ann Bauer
@annbauerwriter

My yoga studio sent me a 'good news!' message yesterday. They're introducing vaccinated-only classes.

I wrote back to cancel my membership w this note:

I am vaccinated but do not participate in vaccine-only events, just as I am white but would not opt into a whites-only class.

10:39 AM · May 25, 2021 · Twitter Web App

Gummi Bear
@gummibear737

Just a reminder that we did this to children

5:46 PM · May 20, 2021 · Twitter for iPad

Emma Woodhouse
@EWoodhouse7

Let me get this straight:

Asymptomatic transmission ISN'T a concern when vaccinated people are asymptomatic, but it IS when asymptomatic people aren't vaccinated?

How's that work? 🤔

(Very well, for the producers of the vaccines)

12:19 PM · May 20, 2021 · Twitter for iPhone

TEAM REALITY

Emma Woodhouse
@EWoodhouse7

Kudos to @Target for good (and small) transitional signage. Daughter & I were the only shoppers in sight who were maskless, but a few staff were as well.

This is progress

[Image of sign: "Face coverings are strongly recommended for guests who are not fully vaccinated."]

4:19 PM · May 20, 2021 · Twitter for iPhone

Bethany S. Mandel ✓
@bethanyshondark

I trust the vaccine I got. That's why I got it. I don't care if other people are vaccinated, because I am.

8:55 AM · May 28, 2021 · Twitter for iPhone

ClownBasket
@ClownBasket

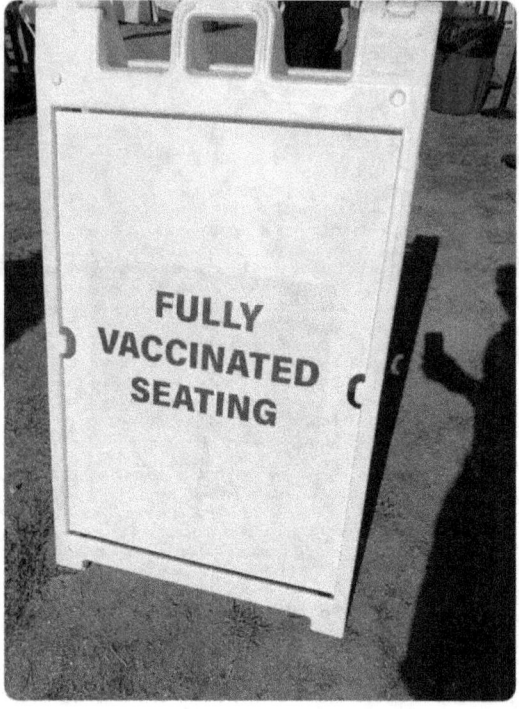

8:27 PM · May 22, 2021 · Twitter for iPhone

Emily Burns 🟢 DMs welcome #TeamReality
@Emily_Burns_V

1/
It's now acknowledged that COVID-19 is primarily transmitted by tiny aerosols generated through NORMAL breathing—not coughing, or flying spit globules.
Based on what we already KNEW about these respiratory aerosols, this should have meant MASKS OUT.

thesmileproject.global/post/airborne-...

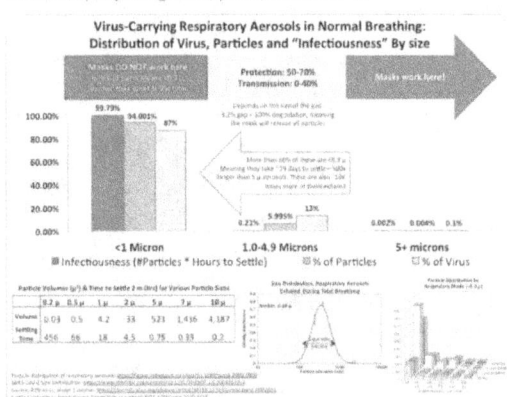

5:04 PM · May 27, 2021 · Twitter Web App

Bethany S. Mandel ✓
@bethanyshondark

COVID has been going on for well over a year. If you still won't let your child go near another child, you are certainly doing psychological damage and it's likely pretty significant. This is a tweet from the playground, where I'm watching it in real-time.

4:44 PM · May 10, 2021 · Twitter for iPhone

M_P
@Reroot_Flyover

Does no one find it odd that it's looking more and more likely that Tony Fauci funded research that led to the Sars-Cov-2 pandemic and also heads our response?

6:53 PM · May 6, 2021 · Twitter Web App

Prof
@covidtweets

Now that the CDC and the WH joined team reality (kind of), seems my days here are numbered.

I want to do a post-mortem on the Spring bump and share a couple of predictions for the Sun Belt before I leave. (1/x)

> **Prof** @covidtweets · Mar 21, 2021
> As I have been predicting, the March-April wave seems to have arrived.
>
> I will elaborate a little bit on what I think is happening here and try to predict what I expect for the next couple of months. (1/x) twitter.com/covidtweets/st...
> Show this thread

11:18 PM · May 13, 2021 · Twitter Web App

I had predicted that our national 7-day cases would peak below 100K, maybe significantly so. The number I had in mind was 80K. We did a little better than that (~73K). One reason is that I expected the NY/NJ area to get a bit worse than they did. ++ I expected the peak to happen during the 1st/2nd weeks of April. We peaked on April 14th. ++ I expected the cases to plummet in the second half of April. By May 1st, 7-day average had dropped by 20K. ++ Now that we are done, let's look forward. What will the Sun Belt wave look like? I am actually optimistic that it will not be too bad. What gives me hope is in the second tweet above. ++ First, an observation. It looks like, overall, there are four periods when there is a strong seasonal stimulus somewhere in the US for the virus spread. See the below chart for details. The numbers represent my rough estimations of the stimulus strength. ++ The strongest stimulus happens during Fall/Winter.

	NORTH	SOUTH
SPRING (3/15-4/15)	Yes (2x)	No
SUMMER (6/15-7/15)	No	Yes (1x)
FALL (10/1-11/15)	Yes (4x)	Kind of (1x)
WINTER (11/15-1/15)	Kind of (1x)	Yes (4x)

11:18 PM · May 13, 2021 · Twitter Web App

Spring is about half as strong and localized in the North. Summer is about a quarter strong and localized in the South. I expect that in the endemic phase, there will no longer be Spring and Summer waves. ++ Why? Because as the % susceptible has dropped like a rock, as a result of past infections and vaccinations, the virus will now need a very strong seasonal stimulus for R to exceed 1. It will no longer happen except for Fall/Winter. ++ This year though, based on this Spring's metrics, I believe our national 7-day average will peak around mid-July with a ~40K 7-day average. Hospitalizations will be much lower but artificially inflated a bit because of the for/with issue. Deaths will barely budge. ++ Finally, at some point, we will have to stop counting cases as they will no longer be a meaningful metric to track. The CNN coronameter will be gone. Soon...

Eric
@The_OtherET

BREAKING: CDC Releases New COVID Guidelines, Says Public No Longer Required to Have Constant Sense of Dread Regardless of Vaccination Status - Experts Call Move "Premature"

8:32 PM · May 22, 2021 · Twitter for iPhone

M_P
@Reroot_Flyover

On covid, how the hell did we go from "man, we're lucky kids aren't really effected by this disease to "we need to keep kids under restriction and in masks for years beyond everyone else until we can foist vaccination on them"? JFC.

9:23 AM · May 28, 2021 · Twitter Web App

TEAM REALITY

Jennifer Sey ✓
@JenniferSey

I am vaccinated. I am not afraid of the unmasked, the unvaccinated or children - inside or outside. I wasn't afraid of them before I was vaxxed.

I will die on a hill to protect bodily autonomy for those who choose not to or can't vaxx & to fight discriminatory passports. 1/2

6:59 PM · May 20, 2021 · Twitter Web App

Being unvaccinated is not the same as having Covid. We can't treat the unvaccinated as if they are toxic waste. That said, I'm not afraid of people with Covid either. 2/2

Emily Burns ✓ DMs welcome #TeamReality
@Emily_Burns_V

3/
And yet, we have this kind of dishonesty from people like Bill Nye. He talks about "stopping the flow of air." Yet he, just as everyone else who has worn a mask, must have felt the air escaping out the top of the mask--fluttering his eye lashes--during this demonstration.

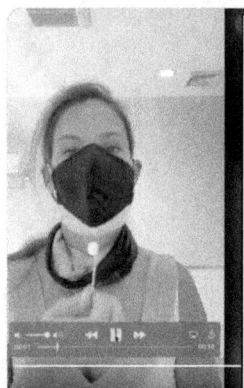

Potentially copyrighted clip showing Bill Nye trying to put a candle out by blowing through the mask, not realizing all the air is going through the gaps so the sides (while his glasses fog).

5:08 PM · May 27, 2021 · Twitter Web App

Eric
@The_OtherET

The chance an asymptomatic kid infects another kid that results in them being hospitalized from COVID is 1 in 42 million.

This is "thousands of times" lower than their risk from bicycles, school buses and swimming pools.

Parents take far greater risks with their kids every day.

11:22 AM · May 19, 2021 · Twitter Web App

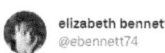
elizabeth bennett
@ebennett74

Seeing a 14-mo old in clinic who hasn't had a checkup since she was 2 mo old (!) because the parents were terrified to take her out of the home. She is OVERDUE for 6 vaccines—all of which are for pathogens much more dangerous to her than COVID. Fear-mongering is a dangerous game.

12:44 PM · May 20, 2021 · Twitter for iPhone

Erich Hartmann
@erichhartmann

So, I'm confused. Is the CDC guidance "gospel" or not?

Because before yesterday it was not to be argued with.

But after changing their tune yesterday re masks... we suddenly have a lot of a-la-carte selective cherry picking going on by States, businesses and institutions 🤷

4:35 PM · May 14, 2021 · Twitter for iPhone

Jenin Younes
@Leftylockdowns1

given the fact the covid-hysteric crowd has just jumped from one type of shaming to the next for the past year (social distancing, masks, now vaccine), I'm beginning to think they just love shaming people.

3:48 PM · May 7, 2021 · Twitter Web App

TEAM REALITY

JUNE 2021

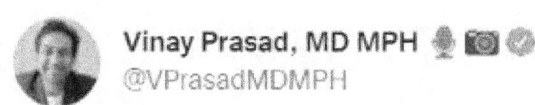
Vinay Prasad, MD MPH
@VPrasadMDMPH

Reaching vaccine targets should be done by persuading adults to get vaccinated, and not lowering the age of vaccination, despite unfavorable safety signals, and net detrimental effects

Pretty simple, actually

4:22 PM · Jun 24, 2021 · Twitter for Android

TEAM REALITY

NYC Angry Mom
@angrybklynmom

The amount of gaslighting in our current discourse is off the charts.

10:24 AM · Jun 29, 2021 · Twitter Web App

"We never had a lockdown". Instead, we had crippling school & business closures, irrational restrictions, based on pseudoscience & you are an immoral piece of shit if you expect to see your family. ++ "Masks prevent spread through droplets", except the virus is primarily spread through aerosols, which we knew from Feb 2020, but masks work on these too, unless you have gaps around the side but nobody ever does, so masks totally work everywhere, especially on kids. ++ "Natural immunity is a myth" unless you get the virus and you have natural immunity, which evidence shows is just as robust as immunity through vaccination, but this is a right wing myth and fuck Rand Paul, a failed eye doctor. ++ "Nobody is suggesting that restrictions should last forever", except that we think that if everyone just wore a mask forever, nobody would get colds or flu and men won't ask women to smile, but nobody is suggesting this. ++ "The government should have just paid people to stay home", except for the ~50% of the population who can't stay home bcs they provide services that are essential to the population & we may have to eat bugs & live in raw sewage if they stayed home. ++ "The virus doesn't discriminate", unless you are over the age of 50, where 95% of the deaths occur or you are under the age of 17, when the virus is more benign than currently circulating viruses, but let's close playgrounds & schools so people take it seriously. ++ "Learning loss is fake", except for when your kid does 1+ yrs of online school & you realize they not only damaged their mental health, but they didn't learn a single thing and need summer school to make up the difference, & that is for kids who actually participated. ++ "Schools aren't closed - only the buildings are closed", except for when your child needs actual school for their mental health, or related services. But don't worry, you can teach your kid to hold a pen through Zoom Occupational Therapy. It's just as good! ++ "It's not Teachers Unions or Democrats keeping schools closed", but just ignore that the school closure map looked just like an electoral map, & that schools all over the US south & GOP areas had a 75% normal school year, & Teachers' Unions donate millions to Dems. ++ "Lockdowns saved lives", except for the places that had the highest excess death rates, where anywhere between 30-50% of deaths were in nursing homes, where people got sick anyway bcs lockdowns don't actually prevent viral transmission when used for 1+ yr, but we didn't have one. ++ "Teen suicides aren't up", except for suicide ideation and ER visits for mental health in teens, as well as overdoses, but nevermind that half of US kids can live normally right this min & yours can't solely bcs of political reasons, not viral outcome. ++ "Protect grandma" so we have to vaccinate everyone, but little kids aren't eligible so they can't get vaxxed yet, even though the virus is less severe in kids than flu & RSV, they have to be vaxxed & masked to protect adults who refuse the existing vaccines available to them. ++ "We have to close beaches, playgrounds, implement outdoor social distancing, & arrest people who have outdoor parties bcs the virus is so dangerous" but only left wing protests don't spread viruses. ++ "We can't have 3 ft of distance, only 6 ft" bcs airborne viruses understand exact measurements of distance, but in fact, I heard that they use the metric system so 1 meter is fine. ++ "COVID came from a wet market" and definitely not from a leak inside of a lab that just happens to exist in the city where it started, and if you suggest this, you are racist because wet markets are totally not part of a stereotype. ++ "COVID is spread through anti-masker Republicans" except for when you look at a map of a big city where COVID is prevalent & it is mostly low income & minority communities who make up the bulk of essential workers bcs we designed a strategy around the safety of the affluent. ++ "We're all in this together", except for the ppl who have huge homes, ability to travel wherever restrictions don't exist while flying 1st class on airlines that let you bypass masks, & UK health secretaries & business travelers who 'contribute to the economy' can skip quarantine ++ "You're an immoral piece of shit if you see your family", unless your family wants to go to French Laundry for my bday, or host a Teachers' Union fundraiser, or travel to the immoral state of Florida during the height of our spike.

Stefan Baral
@sdbaral

If social distancing and masks were primary drivers of the decrease in influenza this year, why did it also disappear in Sweden?

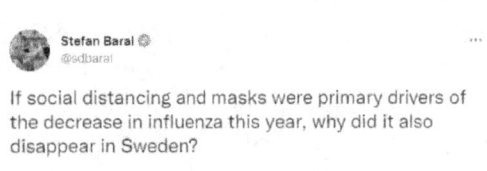
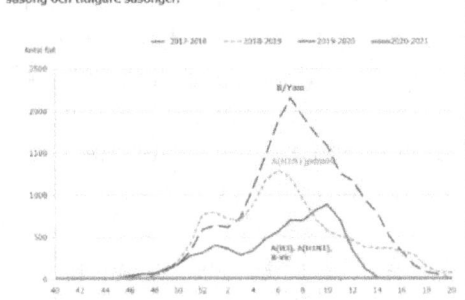

5:08 PM · Jun 27, 2021 · Twitter Web App

Real Developments
@pdubdev

Masks didn't work
Lockdowns didn't work
CDC didnt work
Birx didn't work
Fauci didnt work
Government didn't work

But a healthy immune System did

5:50 PM · Jun 19, 2021 · Twitter Web App

District Al
@districtal

Each day it becomes clear the "crazies" or "Team Reality" or "conspiracy theorists" were actually the ones that were right about childhood vax, natural immunity, lockdowns, seasonality, harm done by school closures and on and on.

7:53 AM · Jun 22, 2021 · Twitter Web App

Victoria Fox
@drvictoriafox

To all the COVID pundits who suggest pre-vaccine, Natural immunity was a completely unviable way to end the pandemic - how else would it have ended? What was YOUR mechanism for ENDING the COVID pandemic without a vaccine OR NI? If you can't answer that - don't post on my feed.

11:43 PM · Jun 18, 2021 · Twitter for iPhone

M_P
@Reroot_Flyover

It's so unfair that Long Covid hits the 25-50 white female demographic so hard. Almost exclusively, really.

9:35 AM · Jun 24, 2021 · Twitter Web App

Vanessa
@vlal42

I don't think the ph/elite media realize what they have done to public trust. I am not a raucous defiant person. I am a rule follower (they used to call me Pollyanna in college). Now I am adverse to any sort of government overreach/safety recommendations. I know I am not alone.

7:08 PM · Jul 27, 2021 · Twitter for iPhone

TEAM REALITY

AJ Kay
@AJKayWriter

Fauci should resign — irrespective of his emails.

Fauci didn't just fail the low-risk people who endured the collateral damage of his policies & reccs.

He failed the at-risk & fearful, too...

11:44 PM · Jun 1, 2021 · Twitter for iPhone

… In a crisis, it is critical that Public Health communicates accurate info about risk & provides strategies and tools that empower people to protect themselves. Fauci offered none of these. Instead, he left people defenseless and terrified... ++ It was cruel to tell people that they would die unless every single stranger they encountered complied w/ draconian restrictions that carried significant individual cost. And it's equally cruel to ask that of others. Anyone reasonable could see that it was impossible... ++ That's like saying, "We can only treat your cancer if someone else takes your chemo." And when that's the only protection you are offered, suddenly the people the unwilling to take your toxic medicine - maybe even people you love — start to look like villains... ++ What if he had said this? "I know you're scared. Pls don't panic. We have this under control. It's what we do. We don't know everything, but we know a lot & we're going to pass what we learn on to you. If you get sick, we will make sure you're taken care of... ++ "…There's going to be sensationalism in the media surrounding this and you need to take it with a grain of salt. In the meantime, here's what you can do to lower your risk…" ++ Instead, Fauci set up a public health platform that essential said, "When you get sick, blame them not me." And, whatever his motivation, this will go down as the most egregious public health failure of our lifetimes… …assuming he isn't allowed to do it again. #FireFauci

---

ClownBasket 🤡
@ClownBasket

My new least favorite thing:

A family together - kids are wearing masks while their parents aren't.

Don't do that.

3:37 PM · Jun 7, 2021 · Twitter for iPhone

Phil Holloway™ 🇺🇸 ⚖️ 🎙️
@PhilHollowayEsq

#Fauci says masks aren't needed unless you're sick and notes the #SARS_CoV_2 is so small it passes easily between #mask fibers #Fauciemails

From: Fauci, Anthony (NIH/NIAID) [E]
Sent: Wed, 5 Feb 2020 03:48:11 +0300
To: Sylvia Burwell
Subject: RE: A couple of quick questions.

Sylvia:
Masks are really for infected people to prevent them from spreading infection to people who are not infected rather than protecting uninfected people from acquiring infection. The typical mask you buy in the drug store is not really effective in keeping out virus, which is small enough to pass through the material. It might, however, provide some slight benefit in keep out gross droplets if someone coughs or sneezes on you. I do not recommend that you wear a mask, particularly since you are going to a very low risk location. Your instincts are correct, money is best spent on medical countermeasures such as diagnostics and vaccines.
Safe travels.
Best regards,
Tony

10:16 PM · Jun 1, 2021 · Twitter for iPhone

Martin Kulldorff @MartinKulldorff · Jun 10, 2021
… and here's the updated, comprehensive list of #SARSCoV2 variants that are resistant to immunity from natural infection

Stinson Norwood
@snorman1776

10 years ago @JenniferNuzzo hosted a panel to discuss pandemic preparedness. Among the panelists: D.A. Henderson, credited with leading charge to eradicate smallpox. Jennifer asks, "what would you do if there were a significant new disease." SARS-like 😷

youtu.be/8rEV857R0LE?t=...

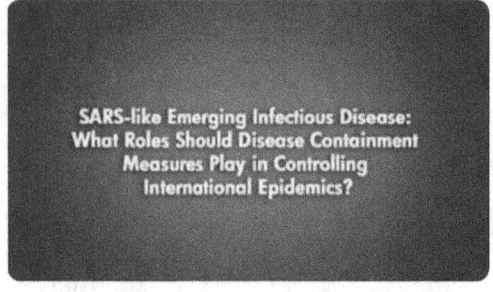

SARS-like Emerging Infectious Disease: What Roles Should Disease Containment Measures Play in Controlling International Epidemics?

2:16 PM · Jun 28, 2021 · Twitter for iPhone

DA Henderson - I've been in lots of epidemics. First thing a poltician does is tell you everything is OK which everyone knows is patently false (laughter) ++ Ronald St John, former Canadian official, formerly with WHO: public will demand response, government will put some of these in place but it's a two-edge sword bc government will say they are doing something but loses trust when it doesn't kep virus out ++ RSJ-let's assume it's like Avian Flu with 50-70% mortality rate, government's response would be radical---this is the "close the border" situation....goes on to say he questions efficacy of closing the borders ++ RSJ - quaranting large #s of people. references 7,000-8,000 as "large" speaks to practical issues, says problems were "unanticipated" and social structure is not there to support quarantining large numbers of people. oh. ++ DA Henderson - still waiting for an example where we could have stopped the first case of an outbreak ++ DA Henderson - what do you do to let public know you're doing something. notes Asia wearing masks. Chuckling, "I think we all agree they don't do a whole lot." ++ DA Henderson - You'd like to tell public something to do. Wash their hands and cough into their sleeve. Notes this seems it would work intuitively but it doesn't seem as effective as you would think. Sort of like mechanistic plausibility? ++ James Blumenstock of Association of State and Territorial Health Officials: when implementing NPIs better have end-game strategy to get out ++ Marcelle Layton, NYC Department of Health: "We don't need to count every case. We almost get caught up, the media gets caught up in how many cases, how many deaths." [Link to conference: https://www.centerforhealthsecurity.org/our-work/events-archive/2010_h1n1_experience/multimedia/index.html]

Matt, Pre-School Diploma
@statomattic

I know these things aren't as socially damaging as masks and distancing, so don't get as much attention, but...

Why in the world are businesses still doing - even advertising - "contact-free" delivery, excessive wiping down surfaces, etc? We know COVID doesn't spread that way.

8:06 AM · Jun 11, 2021 · Twitter for Android

Emma Woodhouse
@EWoodhouse7

CDC now using a social media poll to gather results it will no doubt weaponize against the public, in an ongoing effort to ignore decades of both its own advice & research/data re: natural immunity.

Par for the course.

> **CDC** @CDCgov · Jun 25, 2021
> If you've already had #COVID19 and recovered, you should still get vaccinated against COVID-19.
> Show this poll

2:17 PM · Jun 25, 2021 · Twitter for iPhone

> **Prof Francois Balloux** @BallouxFrancois
>
> Child covid19 vaccination feels like the mother of all ethical minefields. I expected those discussions wouldn't be easy, but the non-trivial rate of myocarditis in teenage boys following mRNA vaccines makes the situation worse than I had anticipated.
> 1/
>
> > **Prof Francois Balloux** @BallouxFrancois · Dec 2, 2020
> > I worry that once those most at-risk will have been vaccinated, the bitter discussions about the ethics of 'herd immunity' strategies may be back on the table. I don't dare to envision how toxic the discussion about vaccinating children for #COVID19 could be.
> > 8/
> > Show this thread
>
> 8:56 PM · Jun 23, 2021 · Twitter Web App

It is now clear that the mRNA vaccines can cause myocarditis in children and young adults. This side effect is most common in teenage boys with a rate of ~0.005%. ++ This has to be balanced against the risk associated with covid19 infection. This is not easy but for example the study below reported a risk of hospitalisation peaking at ~0.002%, but with 70% of hospitalised teenagers having pre-existing conditions. ++ Thus, for healthy teenage boys, the direct cost benefit balance might be neutral or even negative. This is unlikely to be the case for girls or boys with risk factors. What a mess 😐 ! ++ One argument may be that healthy teenage boys should still be vaccinated for the 'greater good', even if this may come at a risk to themselves. Alternatively, one may argue that they would benefit indirectly since they could resume their education (ie open schools). ++ I would personally not feel comfortable with any segment of the population being vaccinated if this meant they might incur a net health risk deficit, even if this benefitted society at large. I'm also not aware of any precedent for such rationale. ++ I wish to stress that the risks associated to both covid19 infection and negative side effects of vaccines remain minuscule in children and teenagers. ++ I don't believe there's necessarily a right or wrong answer to the question whether teenagers should be vaccinated, and there remains considerable uncertainties over exact numbers. That said, I believe regulatory bodies were right to tread very cautiously here.

> **Pajamas It Is** @HeckofaLiberal
>
> I'm going to lose between 1 and 40lbs by next Wednesday.
>
> > **CDC** @CDCgov · Jun 17, 2021
> > As of June 14, national forecasts predict 18,000-140,000 new #COVID19 cases will likely be reported during the week ending July 10. More: bit.ly/CDC_CForecast.
> > Show this thread
>
>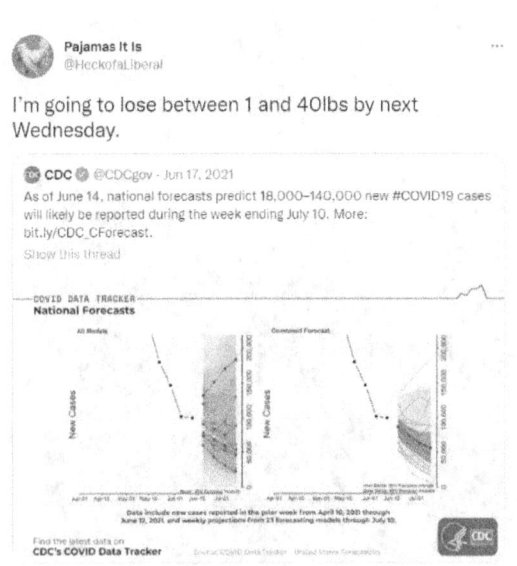
>
> 11:25 AM · Jun 18, 2021 · Twitter for iPhone

> **Craig** @TheLawyerCraig
>
> Lack of nuance is why I started tracking & posting Covid data. Worse is the lack of any desire for nuance. And worst of all is the actual striving for a lack of nuance.
>
> Those in the last category above induce clicks and turn profits off useful idiots in the second-to-last.
>
> 11:49 AM · Jun 16, 2021 · Twitter Web App

> **Wes Pegden** @WesPegden
>
> It's one of those days where a group of experts is meeting at the CDC and most of the random hot takes on the internet claiming they're saying total BS using fudged numbers basically check out.
>
> Implications for the issue at hand aside, it's a step back for science and medicine.
>
> 3:48 PM · Jun 23, 2021 · Twitter Web App

TEAM REALITY

John Ziegler ✓
@Zigmanfreud

Congratulations to the people of Sweden who have allowed their country to, ironically and in direct defiance of the "experts," allowed their country to be one of the very first to effectively achieve "Zero COVID."
#HerdImmunity

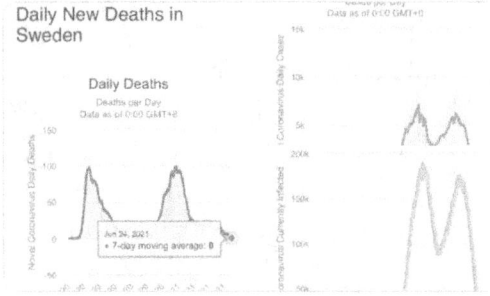

11:21 AM · Jun 25, 2021 · Twitter for iPhone

Tracy Hoeg, MD, PhD
@TracyBethHoeg

Just in case you thought California's school-related policies could not get any less sensical...
Required masks and screening testing even for *vaccinated* kids? 😷😷
What *is* the endgame here?
Do people truly not see how wasteful and pointless this is?

2:08 AM · Jun 22, 2021 · Twitter for iPhone

John Ziegler ✓
@Zigmanfreud

Watching young children on the park playground in masks in a county with 0 real COVID & where the vast majority are vaccinated makes me both sad & angry...

Why is this not the reaction of the vast majority of rational adults?!

When did flaunting child abuse become fashionable?!

3:33 PM · Jun 7, 2021 · Twitter for iPhone

Mark Changizi
@MarkChangizi

At the start of COVID19 hysteria, I excused regular folk. We can't expect everyone to dive down into the data themselves.

But I'm not so forgiving now.

If the state demands suspension of civil rights en masse and crashes the economy, one has a responsibility to question it.

4:18 PM · Jun 17, 2020 · Twitter Web App

Dr Clare Craig (not one of her impersonators)
@ClareCraigPath

Remember that, according to the Pfizer data, you would need to vaccinate 62 children to avoid one case. That equates to 15.5 million to avoid one death of a healthy child. You have to assume that all of those 15.5 million would catch COVID in order to see that benefit.

11:35 AM · Jun 17, 2021 · Twitter Web App

Woke Zombie ✓
@AWokeZombie

Ask yourself: Why hasn't there been one mask study on children in the US the last 15 months? Why is #RationalGround the first to undertake even looking at masks on kids?

8:44 AM · Jun 16, 2021 · Twitter for iPhone

Eli Klein
@TheEliKlein

I was just going into dinner at Employees Only in NYC & the staff at the front insisted I put a mask on in order to walk a few feet to my table

I explained why I wouldn't, then took my business to The Leroy House. It seems great here and all the staff are pleasantly unmasked

6:46 PM · Jun 26, 2021 from Manhattan, NY · Twitter for iPhone

Vinay Prasad, MD MPH 🎙️📺 ✓
@VPrasadMDMPH

The current CDC guidelines are so poor they would recommend a 15 year old boy who recovered from documented covid19 and who got pericarditis from dose 1 go on to get dose 2.

Can we pause a minute to contemplate how staggeringly negligent that is?

7:38 PM · Jun 23, 2021 · Twitter for Android

Justin Hart
@justin_hart

We've lost our ever-lovin' minds.

6:38 PM · Jun 25, 2021 · Twitter for iPhone

TEAM REALITY

Emma Woodhouse
@EWoodhouse7

Hard for me to get tired of reminding people what the Governor of Illinois said last summer.

Turning citizens against each another was - and is - key strategy of the Covid-response campaign.

And it worked.

> **Governor JB Pritzker**
> @GovPritzker
>
> The enemy is not your mask.
>
> If you're not wearing a mask in public, you're endangering everyone around you, so the enemy is you.
>
> 2:00 PM · 7/22/20 · Twitter Web App
>
> **176** Retweets **76** Quote Tweets
>
> **866** Likes

1:40 PM · Jun 28, 2021 · Twitter for iPhone

Eric
@The_OtherET

If you're in a wheelchair you aren't allowed to use a drinking fountain at the Denver airport in order to "reduce the spread of COVID-19".

9:15 PM · Jun 14, 2021 · Twitter for iPhone

Ian Miller
@ianmSC

Has there ever been a policy with less chance of success than "15 days to slow the spread?"

Literally one of the most ineffective policies enacted in modern history, and the people who come up with it are still treated by the media as if they have any clue what they're doing

12:23 PM · Jun 28, 2021 · Twitter for Mac

Gummi Bear
@gummibear737

Gallup Poll of Vaccinated People (Bill Maher - Real Time)

Do you think it is better to try and lead normal lives or stay at home as much as possible?

87% of Republican: Lead Normal Lives

71% of Democrats: Stay At Home

There's alot of cognitive dissonance on the left

7:47 AM · Jun 14, 2021 · Twitter Web App

Emily Burns DMs welcome #TeamReality
@Emily_Burns_V

1/
FL parents send 6 masks to a lab...
- 100% of masks contaminated.
- 50% w/pathogens, including multiple strains of pneumonia- and meningitis-causing bacteria.
- 1/3 with antibiotic-resistant pathogens.
- 2 masks w/more than 70 strains of bacteria.

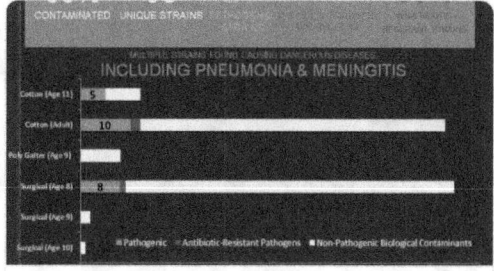

rationalground.com
Dangerous pathogens found on children's face masks
Dangerous pathogens found on children's masks point to the need for a larger, more controlled study

7:42 AM · Jun 16, 2021 · Twitter Web App

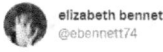
elizabeth bennett
@ebennett74

Last week, I saw an intermittently (passively) suicidal graduating 12th grader in clinic. She told me that she is holding on in hopes of a normal college experience in the fall. There are many kids with thoughts like these. Parents: the time is now to fight like hell for normal!

11:16 PM · Jun 13, 2021 · Twitter for iPhone

TEAM REALITY

Jenin Younes
@Leftylockdowns1

I long considered vaccines among humanity's greatest achievements. I looked forward to the covid vaxx & I assumed I would get it. Before I became eligible, I got covid, so I decided not to. Based on events that transpired, I have since become profoundly skeptical because of: 1/

2:37 PM · Jun 24, 2021 · Twitter Web App

1) the complete disregard for natural immunity; 2) the push to vaccinate everyone, even children to whom the virus poses almost no risk and upon whom the vaxx hasn't been adequately tested; 3) the media and CDC's rush to depict negative events from vaxx as outliers and mild 2/ ++ which is the polar opposite of the approach both the media and CDC took with respect to covid deaths; and 4) mandating vaccines in order to participate in society, attend university, go to work, etc. Call me crazy. I don't think I am. 3/3

Stefan Baral
@sdbaral

My guess is that we would have higher vaccine coverage through positive messaging and meaningful outreach than what has been achieved to date through fear and mandates.

8:26 PM · Jun 24, 2021 · Twitter Web App

Jennifer Sey
@JenniferSey

Science must serve the values of objectivity, discovery, honesty and accountability - all in service of truth.

Science must never betray these values in service of loyalty to a political movement.

Then it's simply not science anymore.

10:35 PM · Jun 15, 2021 · Twitter Web App

Erich Hartmann
@erichhartmann

Florida and Texas and Tennessee are getting millions of refugees from the Blue LockDown States, and they need to launch a "WELCOME!... PLEASE DON'T SCREW UP OUR STATES WITH YOUR DUMBASS POLITICS" campaign, f'n right now. Or we're all screwed.

3:24 PM · Jun 19, 2021 · Twitter Web App

Eric
@The_OtherET

Yahoo needs to explain how, if "lockdown" states are doing better economically than "loose" states, why there are 21 states in the "weak restrictions, strong recovery" quadrant while there are 2 in the "strong restrictions, strong recovery" quadrant...and they barely made the cut

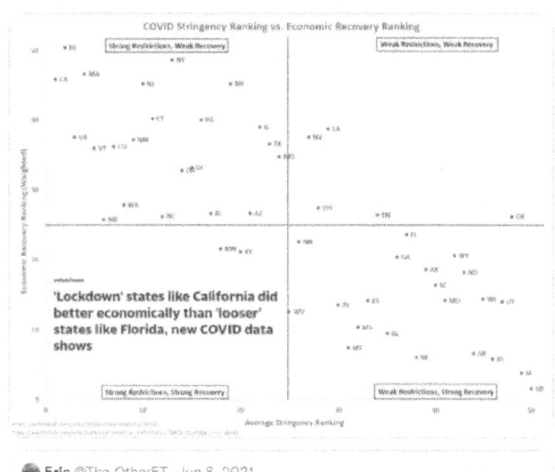

Eric @The_OtherET · Jun 8, 2021
California is currently ranked 44th in leisure and travel recovery and 47th in economy and labor market recovery

What the hell is this article talking about news.yahoo.com/lockdown-state...
Show this thread

5:38 PM · Jun 8, 2021 · Twitter Web App

JULY 2021

elizabeth bennett
@ebennett74

Your daily reminder that all of this was supposed to go away as soon as vulnerable adults had access to the vaccine.

11:19 PM · Jul 19, 2021 · Twitter for iPhone

TEAM REALITY

NYC Angry Mom
@angrybklynmom

Both my husband and I were probably as left-leaning as they come.

I just walked into a room and he was watching Ted Cruz's speech from today, both of us agreeing with every word.

I really hope @TheDemocrats grasp that they are now the party of lockdowns & 'forever pandemic'.

8:58 PM · Jul 29, 2021 · Twitter for Android

I really don't think I can describe how bad the last week has been. Both @CDCgov & @POTUS have shifted the goal posts once again, making it at least another year before we can really have our lives back, as well as our children's. It's not the virus doing this. ++ I can't believe we are approaching 18 months and they think that we should just continue to sacrifice months and years of ours & our children's lives for a virus that has three effective vaccines. We never promised 0 illness. ++ A small child living in a blue area now has no memory of what life was like before the pandemic - freely hugging friends, breathing freely, and socializing without worrying about viral spread. Yet kids in red areas have normalcy. ++ Mark my words, teachers' unions are salivating right now and they can't wait to negotiate more concessions so that they can ensure our children have their third interrupted, miserable school year. How much more should we be expected to take? ++ I want to be clear - my values haven't changed. But I am now coping with Maslow's Hierarchy - basic freedom to exist w/o all aspects of life revolving around a virus, & the constant threat of restriction comes before literally anything else that Dems have to offer.

Virál Myãlgía MD, PhD
@cuntrarian4data

Replying to @Leftylockdowns1

I already have "Twitter-dibs" on omicron being the world ender.

5:38 PM · Jul 6, 2021 · Twitter Web App

ClownBasket
@ClownBasket

BREAKING: Delta variant is now the MOST deadly disease ever imagined.

11:00 PM · Jul 1, 2021 · Twitter for iPhone

Bachman
@ElonBachman

Casual conversation with pediatrician:

Me: "I looked at all 12 mask RCTs, and it seems there's no evidence of efficacy"

Her: "Well I think they work"

Me: "Interesting. Why?"

Her: "I wore a mask this year and never got the flu"

Science delenda est

12:09 PM · Jul 18, 2021 · Twitter Web App

Ian Miller @ianmSC · Jul 27, 2021

It's absolutely criminal that anyone in this country still takes the CDC seriously when cases in Michigan went down -98.2% after the head of the agency told Michigan to "shut things down" and "reimpose restrictions" in April and they ignored her

They're a complete and utter joke

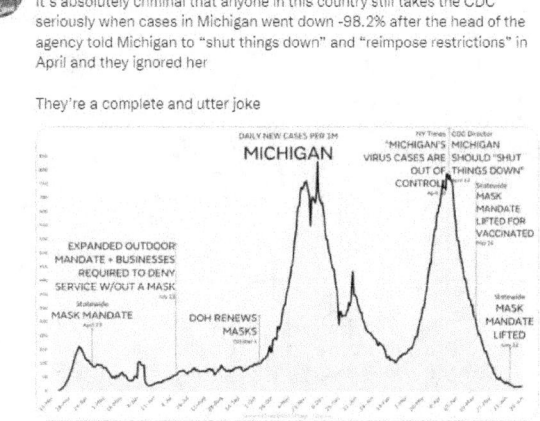

Prof Francois Balloux
@BallouxFrancois

If a car is repaired after a collision, it will have a higher risk of mechanical failure afterwards. 'Long-accident' issues are far more likely to arise if the car suffered severe damage rather than just a minor dent. The same is expected to be generally true for viral infection.

6:26 AM · Jul 23, 2021 · Twitter Web App

TEAM REALITY

Bachman @ElonBachman · Jul 19, 2021
What about a fox, a chicken, and a sack of grain?

elizabeth bennett @ebennett74

I have sent more suicidal teens from my office to the ER in the past year than in my entire preceding pediatric career of 20 years. #TrueStory

9:23 PM · Jul 23, 2021 · Twitter for iPhone

Infectious Disease Ethics @ID_ethics

This could be the most important graph of the pandemic:

Is it too late to flatten the curve?

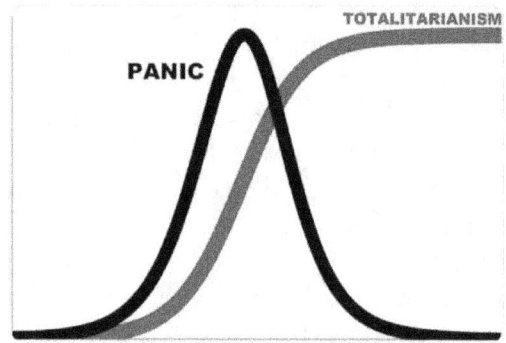

5:57 PM · Jul 25, 2021 · Twitter Web App

Michael Tracey @mtracey

There seems to be an enduring temperamental divide between those who stopped viewing COVID as an acute problem requiring constant attention once vaccines became universally available, and those for whom universal vaccine availability made little or no difference in this regard

12:31 PM · Jul 30, 2021 · Twitter Web App

Erich Hartmann @erichhartmann

Good morning everyone! #MasksDontWork 😷

9:03 AM · Jul 29, 2021 · Twitter Web App

Michael P Senger @MichaelPSenger · Jul 29, 2021

"Are you going to be going suburb to suburb, street to street, door to door, knocking on these and actively looking for people who are in the wrong house and fining them on the spot?"

Sydney police: "Absolutely."

Not a hint of irony...

District Al @districtal

The CDC misrepresents data. Conducts flawed studies. Sets policy based on rejected papers in India and mannequin experiments in a lab. If you follow the CDC you aren't following science. You are following politics.

8:24 AM · Jul 28, 2021 · Twitter Web App

Eric @The_OtherET

The 10 states with the highest COVID hospitalizations per capita 7/24/2020 (left) vs the 10 states with the highest COVID hospitalizations per capita 7/24/2021 (right). I think I'm picking up a pattern here.

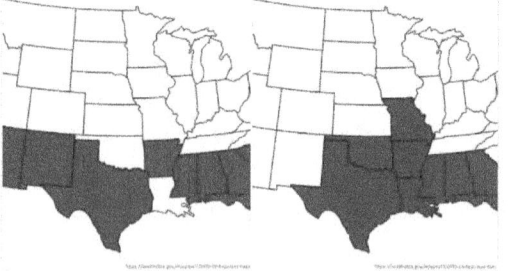

Victoria Fox @drvictoriafox

The oddity that is mask theater: at a restaurant picking up food - a family who didn't have vax cards was forced to wear masks in order to walk to their table. After reaching their table they were allowed to remove their masks because? ..I guess COVID doesn't spread while seated?

9:16 PM · Jul 8, 2021 · Twitter for iPhone

1:51 PM · Jul 24, 2021 · Twitter Web App

TEAM REALITY

Brumby
@the_brumby

Imagine being so stupid you still think this is driven by human behavior / interventions / mandates

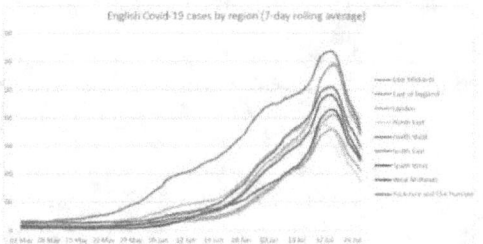

5:16 PM · Jul 29, 2021 · Twitter for iPhone

Mark Changizi
@MarkChangizi

COVID19 was designed to ruin your opinion of literally everyone you once respected.

9:13 AM · Jul 20, 2021 · Twitter Web App

Vinay Prasad, MD MPH
@VPrasadMDMPH

One more point:
I actually think it is ok to mask mandate for 2-3 months without RCT

But 2 to 3 years?
And you never generate data?
And after vaccination?
And for 2 year olds for years?

Are we living in the stone age

12:17 PM · Jul 29, 2021 · Twitter Web App

Eli Klein
@TheEliKlein

The CDC needs to accept the reality that Covid will be endemic so ~everybody will be exposed numerous times throughout life

Most people are better off being exposed to Covid for the 1st time when they're younger & healthier

Policy should be based on this reality, not alarmism

3:32 PM · Jul 26, 2021 · Twitter for iPhone

Ian Miller
@ianmSC

Fauci 5/19: "If you are vaccinated, you can feel safe—that you will not get infected either outdoors or indoors"

Fauci 7/23: "Understandable" to tell people "Good that you're vaccinated, but in a situation where you have people indoors...you should wear a mask"

The Science™

1:46 PM · Jul 24, 2021 · Twitter for Mac

Stefan Baral
@sdbaral

Just a FYI that testing positive for #SARSCoV2 with pre-existing immunity (vaccination/infection) and experiencing mild or no symptoms is a test result, not a disease.

8:58 AM · Jul 18, 2021 · Twitter Web App

Justin Hart
@justin_hart

The fact that the science establishment has outlawed the term "natural immunity" and that you need to refer to it by its new phrase... "prior immunity" should tell you everything about the state of our science today.

12:29 AM · Jul 28, 2021 · Twitter for iPhone

Craig
@TheLawyerCraig

A Toyota Prius full of vaccinated 85-year-olds is more likely to see one of their number hospitalized or die from #Covid19 than a bus full of unvaccinated 25-year-olds. You now know why highly vaccinated countries will show bad outcomes increase in the vaccinated over time.

1:14 PM · Jul 21, 2021 · Twitter Web App

Emma Woodhouse
@EWoodhouse7

Isn't it weird that the CDC is more willing to say the vaccine doesn't work than they are to say masks don't work?

9:57 AM · Jul 29, 2021 · Twitter for iPhone

District AI
@districtai

In 15-20 years you will encounter a young adult in their 20s and you will immediately know how their parents handled this pandemic.

6:16 PM · Jul 20, 2021 · Twitter Web App

Woke Zombie
@AWokeZombie

How many billions of dollars did US taxpayers hand over to the CDC for Covid and we STILL don't know how many Hospitalizations are WITH Covid vs. FOR Covid.

Why is that? Where are the studies on that?

10:51 PM · Jul 28, 2021 · Twitter Web App

TEAM REALITY

Woke Zombie ✅
@AWokeZombie

#US #COVID19 The CDC just made 163 million people mask up because of this:

There was no citation, no reference, no science or data included in the claim.

[image of text excerpt: "data from recent outbreak inve... ns of the virus thats cause Co... icates that in rare occasions, gious and may spread the virus... ly warrants an update to our r..."]

11:02 PM · Jul 27, 2021 · Twitter Web App

Prof Francois Balloux ✅
@BallouxFrancois

We're entering an interesting stage of the pandemic, with most of Europe and the US well advanced into an 'endemic transition'. I see light at the end of the pandemic tunnel, that some may choose to dismiss. I predict that discussions will harden even more over the coming months.

8:45 PM · Jul 15, 2021 · Twitter Web App

Mark Changizi
@MarkChangizi

The first two ethical principles for medicine are...
(1) Do no harm.
(2) Obtain consent.

Both were violated.

12:20 PM · Jul 11, 2021 · Twitter Web App

Prof
@covidtweets

Public Health: We need NPIs to stop the spread. It is in our control.

SARS COV2:

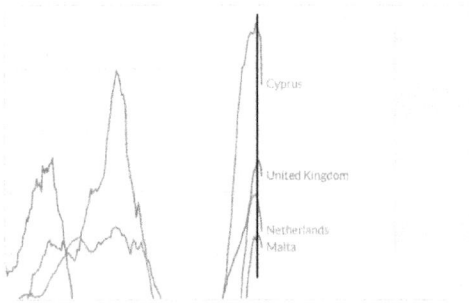

10:12 PM · Jul 24, 2021 · Twitter Web App

Ian Miller
@ianmSC

Cases in Thailand are up 696,400% since National Geographic said that they had done an amazing job preventing COVID from "gaining a foothold" due to a commitment to masking & public health

Nailed it again, The Science™!

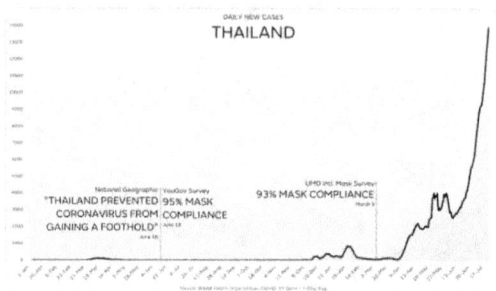

11:59 AM · Jul 27, 2021 · Twitter for Mac

Kyle Lamb
@kylamb8

Airports around Florida are literally breaking records for number of passengers coming in and out of Florida. The scaremongering isn't impacting many who cherish a free and open society where they evaluate their own risk.

7:47 PM · Jul 24, 2021 · Twitter for Android

Emma Woodhouse 🌸
@EWoodhouse7

"Masked & Unmasked Vaccinated People and Masked & Unmasked UnVaccinated People are infecting Masked & Unmasked Vaccinated People and Masked & Unmasked UnVaccinated People.

So get Vaccinated, and wear a mask."

Did I get that right, @CDCDirector?

4:18 PM · Jul 27, 2021 · Twitter for iPhone

Virál Myãlgía MD, PhD
@contrarian4data

On my night shift tonight, I convinced 2 obese 50-somethings to vaccinate based upon an individual risk discussion.

What I didn't do:
1) dismiss their concerns
2) condescend
3) imperiously cite "the science"
4) get political
5) tell them to do it for others
6) wear a bow tie

8:08 AM · Jul 23, 2021 · Twitter for iPhone

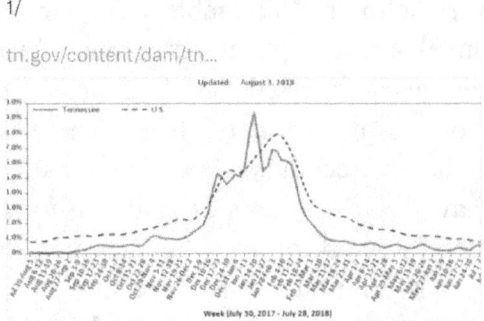

 Josh Stevenson
@ifihadastick

This is a FLU epicurve. It was happening in 2017/18. There were cases cases cases. You likely knew it was flu season, and accepted that reality. We knew it killed people. We have a shot for it. We even had hospital surges because of it.

1/

tn.gov/content/dam/tn...

12:32 AM · Jul 15, 2021 · Twitter Web App

This is the seasonal curve of the other 4 Coronavirures, which are endemic. You lived your whole life and never knew their patterns, prevalence, seasons, or names. Why do you think they broke these out by geographic regions? ++ The important thing for the "case" curve is that we see it decouple from hospitalizations and deaths. How do we do that? Stop wasting finite resources and diminishing public trust on the population LEAST AT RISK like already immune and children. ++ The public is starting to turn on Public Health. Trust has evaporated. The only people taking them seriously are already vaccinated and still wearing masks. When PH tries to convince immune people that they need a vaccine, they are wasting 2 precious resources: ++ Attention, and credibility. We should be hearing about age-stratified risk factors, medical conditions, etc. If you are still at risk, the vaccine is incredibly effective. PH should target those specific people- not kids, or previously immune. ++ Back to the Cases vs hospitalizations What we want to see is an ever diminishing hospitalization rate, and diminishing mortality rate, due to increased population immunity. ++ But our media and hysteria has us fixated on contextless concepts such as "surges" fed by reporters and Social media attention hogs who will desperately cling to the attention they had last year by acting as if these cases are just as significant as they used to be. ++ Inevitable Media interpretation of The Chart on Left: "Hospitalizations Spiking as Delta Variant increases!" Chart On the Right- shows total available beds, total inpatients, and how many are Covid- Will get ZERO Clicks. ++ Most doctors want to truly help their patients, and I expect that many are rightly trying to cut through the political rhetoric (they are probably also not on twitter) and provide real, nuanced, advice. When Public Health leaders contribute to the media circus. ++ And point fingers at the "anti-vax" or "science deniers" or "rural whites"- they only exacerbate the problem that some people who are at risk still need to take them seriously. ++ Case in point: When people like Michelle Fiscus target "white, rural conservatives" as the primary culprits for hesitance. What does that accomplish? By the way, look at the comparison of % vaccinated by Race. ++ So, in summary- the "Public" in Public Health has largely already decided what they're going to do about Covid. "Cases" don't mean what they used to. PH can still protect those still at risk, but they are going to have to stop wasting our time.

TEAM REALITY

Wes Pegden
@WesPegden

I am surprised that more people in public health are apparently not disturbed by the ethical implications of requiring young people who have confirmed prior COVID-19 infection to receive COVID-19 vaccines to be allowed to get a college education.

1/

3:35 PM · Jul 13, 2021 · Twitter Web App

There is now extensive and consistent real world data showing the excellent protection provided by immunity resulting from infection. The tradeoff of vaccine benefits vs possible risks—overwhelming favorable in most adults—are completely different when previously infected... ++ ... and the absolute benefit to the broader community is orders of magnitude lower. There is plenty of space for disagreement on whether and how to recommend vaccination after prior infection. But for mandates, we should have high standards of evidence for a large benefit. ++ I'm not aware of a single scientific work of any kind claiming to analyze tradeoffs inherent in vaccine mandates for 18-year olds with prior COVID-19 infection. But these mandates will be the norm for many such people, and Public Health (and the @CDC) is choosing silence. ++ On the whole, the unfavorable impression this leaves is that decisions about mandates may be driven less by a desire to achieve collective good through careful decision-making than by a dark human tendency to enforce one's will on people perceived to be dangerous or inferior.

Hold2
@Hold2LLC

I need help understanding something.

What is the goal with CoV2? At what point is the situation "good enough"?

Today, how does CoV2 burden on hosps/mortality compare to other viruses? Those wax and wane naturally; shouldn't CoV2?

@shvetaraju @VPrasadMDMPH @BallouxFrancois

/1

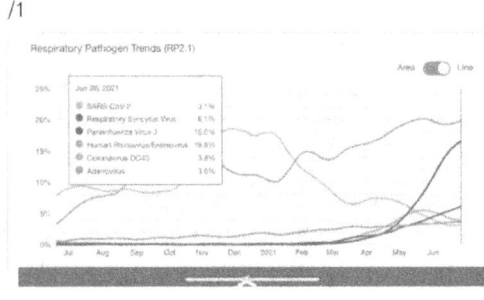

2:10 PM · Jul 11, 2021 · Twitter for iPhone

Angry Cardiologist
@AngryCardio

I know it is hard to believe, but not everyone fears illness.

8:35 AM · Jul 23, 2021 · Twitter for iPhone

AJ Kay
@AJKayWriter

If you post a 20-tweet "take-down" on the JAMA CO2 mask letter but were silent on the CDC mannequin simulation or the hairdresser case study claiming to show efficacy (or any of the other seriously flawed mask papers) you cannot claim your criticism is rooted in science...

10:33 AM · Jul 3, 2021 · Twitter Web App

Ann Bauer
@annbauerwriter

If you're a person who supported keeping schools closed for >1 year and now wants children masked & distanced, even while adults are packing into bars, you never get to say "But we didn't know..." We know. This is abuse. It's political. It's about money & power. Zero forgiveness.

12:25 AM · Jul 10, 2021 · Twitter Web App

Prof Francois Balloux ✓
@BallouxFrancois

Two weeks on, and so far both my predictions that the tweet below would annoy many people, and that 'freedom day' would only have a marginal effect on the dynamic of the epidemic seem to hold ...

Prof Francois Balloux ✓ @BallouxFrancois · Jul 12, 2021
At the risk of annoying many people, I'll go out on a limb by suggesting that the relaxation of the remaining COVID19 measures on the UK on July 19th might only have a marginal effect on the dynamic of the epidemic.

12:01 PM · Jul 26, 2021 · Twitter Web App

TEAM REALITY

 Anish Koka @anish_koka · Jul 11, 2021
Always worth remembering that in the midst of a once in a century pandemic our public health officials almost unanimously advocated to delay the vaccine because of politics.

60 even signed a direct letter to the CEO of Pfizer demanding a delay.

 **Prof**
@covidtweets

Masking is already up in blue areas. No state/city mandate is needed.

Red areas will ignore whatever the CDC says.

CDC now recommending that vaccinated people also wear masks would accomplish absolutely nothing other than sending the message that vaccines don't work.

1:06 AM · Jul 27, 2021 · Twitter Web App

 **Gummi Bear**
@gummibear737

This data = no such thing as herd immunity for C19

It's a respiratory virus...it mutates every season

Vaccines give you short lived immunity vs infection

Hopefully long term protection vs disease

You need to treat this like the flu

Flu/C19 shots for high risk

Stop testing

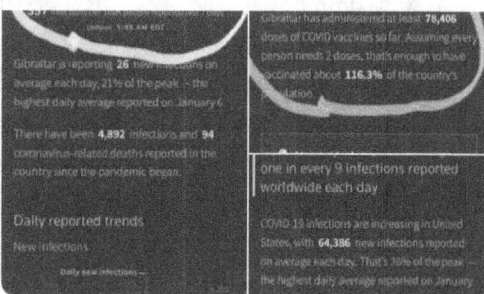

9:37 PM · Jul 18, 2021 · Twitter for iPad

 **Emma Woodhouse**
@EWoodhouse7

Where's the public health honesty about the BMI of young adults who are being hospitalized with Covid?

No shaming necessary, but it helps exactly no one when officials won't say out loud that obesity is a significant risk factor.

6:46 PM · Jul 25, 2021 · Twitter for iPhone

 **Bethany S. Mandel**
@bethanyshondark

Anyone else loose faith in medical providers that do this?

9:47 AM · Jul 26, 2021 · Twitter for iPhone

 Emily Burns DMs welcome #TeamReality
@Emily_Burns_V

All that needs to happen for us to be done is for the CDC to say:

"It's OK if you get infected. The vaxes protect you from severe disease. Sure, you might get the virus, you might pass it, but everyone who wants to be protected is."

The CDC has embraced COVIDZero, w/a leaky vax

> **Emily Burns** DMs welcome #TeamReality @Emily_Burns_V · Jul 29, 2021
> This attempt to blame spread on unvaxxed when a place like Gibraltar w/100%+ adults vaxed, but w/4x the cases per capita as us is really wearing thin.
>
> It's not a sterilizing vax. We need to update expectations accordingly, and stop this divisiveness.
>
> graphics.reuters.com/world-coronavi...

10:48 AM · Jul 29, 2021 · Twitter Web App

Prof
@covidtweets

I have some thoughts to share regarding delta, and the current rises we are seeing in Europe and the US south.

I will also share my predictions about the coming weeks.

Before proceeding, please read the quoted tweet and the whole thread it was in. (1/x)

Now that you have completed your reading assignment, let's begin - sorry, occupational habit :) As many of us here have observed, COVID is following its pattern from 2020 with the same inflection points but smaller waves. However, the Sun Belt wave came a bit later. ++ Earlier I had been predicting a widespread Sun Belt wave (see the thread above), then I revised my prediction to a group of localized outbreaks. The reason for this was because the Sun Belt wave had not started when it did last year. ++ Then delta came. The evidence so far suggests while it is not causing more severe illness, it appears to be more transmissible. See the chart below. Now multiply the numbers on the left with 2-3, depending on how much more transmissible delta is. ++ Imagine a city with 30% susceptible and 50% seasonal stimulus. Find that spot on the chart. Please note that these numbers are only for illustrative purposes. As you see, the R0 is slightly below 1.0 with the given parameters. When delta doubles the numbers on the left? ++ I believe this is what is happening in the south and elsewhere. Places that would normally not see an increase are now seeing a wave

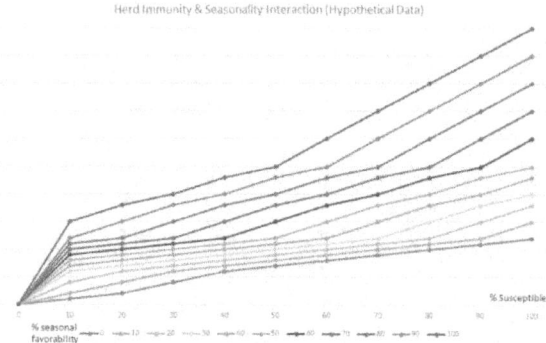

with delta. Because delta became dominant a bit after the inflection point, it created a delayed Sun Belt wave. ++ What is next? Last year, the south wave peaked right around now, meaning it reached R0 = 1.0 as a result of declining seasonal favorability and % susceptible. There are two possibilities - the wave may run its 8-10 weeks course like past waves, or may peak soon. ++ Which one happens will be determined by the interaction between seasonal favorability and % susceptible and where/when it arrives at 1.0 with delta. I would attempt an estimate, but I don't have the data (or time) to do it. ++ However, the good news is that none of this matters. I am not seeing a plausible path to another lockdown in the US at this point, maybe except for the most fanatic states like CA. Even that is highly unlikely. Why? ++ The governors see that: 1. There is no will in the populace anymore. 2. Lockdowns are not really that effective. 3. They have their exit tickets from the situation they had cornered themselves in - vaccines. 4. The social/economic harm will take them down if they insist. ++ Simply, they have no reason to reinstate the same restrictions, and have every reason to avoid it. Even the UK is not doing it. Besides, I suspect they also fear the backlash from those who are vaccinated, who would no doubt see it as the non-vaccinated people's fault. ++ Even more importantly, they see that if they do a lockdown, the people they are trying to convince to take the vaccine will have another argument against it - why take it when it will not change anything? This is probably why we haven't seen mask mandates come back yet. ++ The most that can happen, as I see it, is a small possibility that mask mandates may come back in a few states with blue governors. That's it. Otherwise, life has now resumed and will not be interrupted for this virus again.

> **Emily Burns** 🗸 DMs welcome #TeamReality
> @Emily_Burns_V
>
> 1/
> Trying to mask the abject failure of blue state COVID responses, a new success metric has been rolled out: Vax levels. But whole pop. vax levels, mask much lower variance in at-risk groups. What's more, deaths in 65+ from Jan-Jun are NOT linked to vax levels.
>
>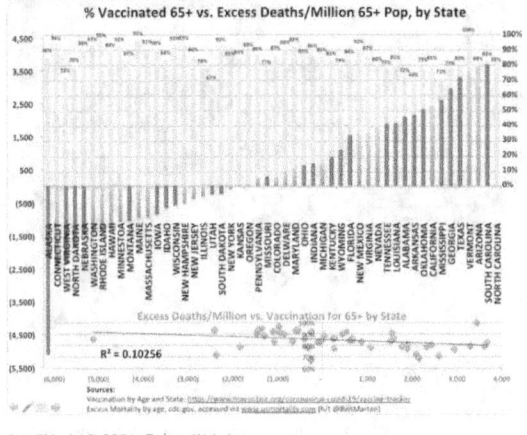
>
> 3:11 PM · Jul 8, 2021 · Twitter Web App

Some will say that the vaccination level of the population is important b/c those other vaccinations are shielding the at-risk further. But excess deaths since January are NOT tied to higher levels of whole population vaccination. ++ The push to vax well beyond the at-risk represents another installment of "following the science" where sadly, fealty to "the science" does not produce any measurable result. This new metric—whole pop. Vax—is designed to wash away all the other failures ++ "The Science" is not concerned with results, either COVID deaths, or any other collateral damage from their policies. This is why, following the science has no impact on COVID outcomes, and highly negative outcomes for everything else. ++ High vaccination levels beyond those who need protection is meant to absolve these states from their gross failures. It should not. These policies have been catastrophic. Their architects and adherents should be held to account. ++ It is interesting to note that there is much less variation by state in vaccination levels in older groups. It is specifically in younger groups where the greatest polarization occurs--with a 10x difference between least and most. ++ This represents just one more example of "The Science" being willing to sacrifice children to adult fear. Masking is tightly tied to child vaccination levels. As are adult fear of COVID, and low levels of in-person education. ++ What this graph shows, is not that there is no benefit to vaccines, it shows that there does not appear to be much—if any—benefit to vaccinating those who are not at-risk. ++ Even in older groups, people without co-moribidities rarely die of COVID-19. Age is basically a proxy for co-morbidities. Remember, the average age of death for COVID is above the national average, and the average number of co-morbidities is 4.

> **Eric**
> @The_OtherET
>
> Here's what will happen next:
>
> 1. Some cities/counties/states will mandate masks again citing CDC guidance, while others will not
> 2. COVID metrics in surge areas will decrease at roughly the same time regardless of mask policy
> 3. Experts claim "victory" for masks anyway
> 4. Repeat
>
> 3:48 PM · Jul 27, 2021 · Twitter Web App

> **elizabeth bennett**
> @ebennett74
>
> China convinced Italy to lock down, then the whole West decided it was a great idea. Now we are stuck in a politicized, media-driven, fear-fueled monomaniacal groupthink. As a non-political doctor with basic numeracy, this whole thing has been so confusing. Why aren't we done?
>
> 10:52 AM · Jul 20, 2021 · Twitter for iPhone

TEAM REALITY

 **Eric** @The_OtherET

"The new COVID variant is SUPER airborne and SUPER infectious so we are asking everyone to start wearing that porous, loose-fitting cotton mask they got at the convenience store for 99 cents and haven't washed in 5 months"

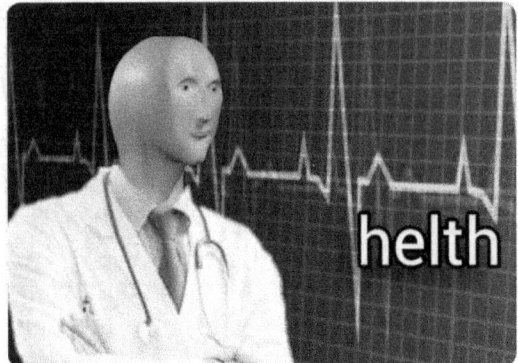

11:58 PM · Jul 27, 2021 · Twitter for iPhone

 **Woke Zombie** @AWokeZombie

I'm so tired of public figures lying to us about Covid.

9:08 AM · Jul 28, 2021 · Twitter for iPhone

 **Jenin Younes** @Leftylockdowns1

Can the people who want to live in a perpetual state of disease panic please just leave the rest of us out of it? If you want to stay home forever and wrap yourself in gauze be my guest. I'm not doing it, ever again. I have one life and I'm making the best of it.

11:39 AM · Jul 17, 2021 · Twitter for iPhone

 **Zac Bissonnette** @ZacBissonnette

You don't have to be a covid truther or a Republican or anti-science to think that the CDC, the FDA, the WHO, and, yes, Dr. Fauci have turned in a performance over the past year and a half that is just spectacularly unimpressive. None of this is confidence inspiring.

3:12 PM · Jul 28, 2021 · Twitter for iPhone

 **Michael P Senger** @MichaelPSenger

DeSantis: "I think it's very important that we say unequivocally: no to lockdowns, no to school closures, no to restrictions, and no mandates...Americans should be free to choose how they govern their affairs...not consigned to live in a Faucian dystopia."

From Evan Donovan

2:51 PM · Jul 29, 2021 · Twitter Web App

 **Clifton Duncan: Good Looking Loser.** @cliftonaduncan

Mandating a BMI of less than 25 would be incredibly effective in combating severe Covid, but it would be wrong to rob someone of the choice of what they do with their body

11:00 AM · Jul 31, 2021 · Buffer

 **Erich Hartmann** @erichhartmann

There was never a US hospital system "overloaded" with COVID cases. Ever.

Why do we want to pretend there was?

4:49 PM · Jul 29, 2021 · Twitter Web App

 **Jennifer Sey** @JenniferSey

I'm so dumbstruck by this view to deny healthcare to the unvaccinated. I'm repeating myself I know but I'm honestly dumbfounded ... would anyone advocate denying HIV medications to a person who becomes HIV positive from unprotected sex? Real question.

5:31 PM · Jul 24, 2021 · Twitter Web App

 **Jennifer Sey** @JenniferSey

To be clear I'm not actually concerned this will happen (not really) I'm just saddened that so many seem to really want it.

7:44 PM · Jul 24, 2021 · Twitter for iPhone

TEAM REALITY

AUGUST 2021

Clifton Duncan: Good Looking Loser.
@cliftonaduncan

Zero Covid is cool as long you're fine with Zero Society

12:16 PM · Aug 15, 2021 · Twitter for Android

TEAM REALITY

Prof @covidtweets · Aug 29, 2021
Almost every single death from COVID (except for very old and frail) happening now is preventable.

The fact that it is still happening at these levels is a massive failure of public health.

"Experts", all of these deaths are on you... You failed... (1/x)

You failed when you took people's livelihoods, students' education/friends, the joy out of life, away for months, based on iffy models with weak parameters and assumptions. You failed when you never admitted that it was unnecessary. We the people took notice... ++ You failed when you never acknowledged that this was a mild illness for most people. You failed when you closed playgrounds and banned beaches, based on no data at all, despite them being among the healthiest and safest things to do... We watched... ++ You failed when you never applied a trade-off analysis to the actions you recommended, and just because you were happy working from home with Zoom, assumed that everyone could follow your guidelines. While saying things like "we are all in this together." We listened... ++ You failed when you acted like shameless hypocrites and praised both the guy with the grim reaper costume on Florida beaches and those who had massive gatherings, simultaneously, in May/June 2020. We paid attention... ++ You failed when you blamed every single red state governor for rising cases but never said/wrote anything criticizing Newsom last winter. You failed when you praised Cuomo, despite his massive failure causing thousands of deaths in nursing homes. We waited... ++ You failed when you ignored WHO, all of Europe, and almost every other country in the world who are not masking kids. You failed when you never entertained the possibility that quarantines were unnecessary and school was too essential to sacrifice. We got mad. ++ You failed when you first said masks were not necessary, then said cloth face coverings were absolutely essential, then said one is not enough - use two, then said use surgical masks, then said use N95... We couldn't believe how incompetent you were... ++ You failed when you never explained how the virus was both airborne and a piece of cloth would stop it, You failed when you never explained how states with draconian measures were having the exact same curves as those with, We began ignoring you... ++ Now it is the classic "the boy who cried wolf" situation. Millions of people have stopped trusting you long ago. You have zero credibility. You are telling them to get vaccinated. Guess what, they are not doing it... ++ Because they are not doing it, a lot of them are dying. Because they are not doing it, you are trying to put vaccine mandates on them. You are forcing useless masks one more time. They don't care. ++ I will tell you what you should do if you really want to save lives and not in this for the spotlight: Stop talking. Stop giving advice. Stop appearing on TV. Stop tweeting. Let PCPs, likely the only medical professionals left that people trust, clean up your mess.

Anish Koka
@anish_koka

Don't care if you're a doctor or a comedian or have a phD in causal inference, if you don't think a well done RCT on ivermectin with hard clinical endpoints would be useful right now... stick to making funny Ortho/anesthesia tik-tok videos

11:39 AM · Aug 29, 2021 · Twitter for iPhone

Prof Francois Balloux
@BallouxFrancois

The suggestion to administrate covid vaccine boosters to everyone, irrespective of age, health and immune status, strikes me as epidemiologically, morally and ethically wrong.

5:18 PM · Aug 19, 2021 · Twitter Web App

TEAM REALITY

David Zweig
@davidzweig

A 3,000 word deep dive into the evidence behind student mask mandates

Many countries around the world - with vax rates, case rates, and mortality above and below the US - do not require masks on students. Why does the US?

My latest for @NYMag

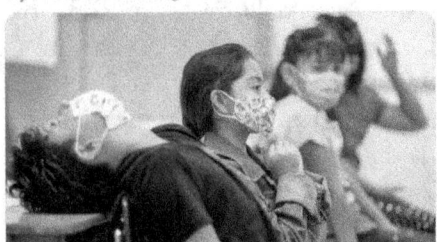

nymag.com
The Science of Masking Kids at School Remains Uncertain
A large, groundbreaking study suggesting no clear benefit from school mask mandates has many experts questioning the policy.

1:53 PM · Aug 20, 2021 · Twitter Web App

David M
@ComradeDoom1

It's been humbling to realize that wacko rightwingers in Huntington Beach, bikers in Sturgis, and dumb college kids on spring break were basically right about covid while 90% of the people I respected pre-2020 were disastrously wrong.

10:12 AM · Aug 9, 2021 · Twitter Web App

BOUTROS 木
@boutros555

Australia is literally tearing itself apart because its leaders are in a state of denial about the fact that good behavior doesn't stop respiratory viruses.

Let that sink in.

10:42 PM · Aug 21, 2021 · Twitter Web App

District AI
@districtai

We have been trying to defeat Air and Death for 18 months. I think we can call this one. We lost. RVs spread. We cant stop them. We can simply make our own personal choices in choosing individually what we to partake in. Thats it. Live life. See you in 10 drinks!

7:14 PM · Aug 27, 2021 · Twitter Web App

Mark Changizi
@MarkChangizi

If you find yourself supporting the removal of the unvaccinated from society, then you're now part of a special club that includes history's favorites such as the Nazis, Chinese Cultural Revolutionaries, Islamic revolutionaries, Hutus, and just SO many glorious others!

1:16 AM · Aug 6, 2021 · Twitter Web App

David Zweig
@davidzweig

If you are a vaccinated adult and living a relatively normal life, but find waiting for a pediatric vax "intolerable," please understand that your unvaccinated child is at a significantly lower risk than you

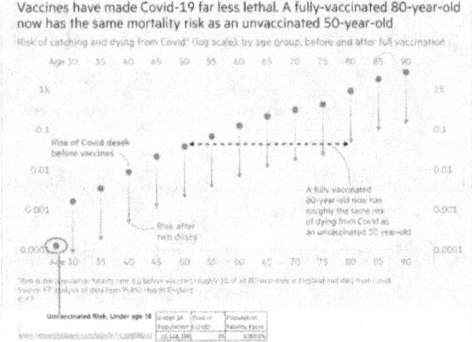

👤 Phil Kerpen

10:26 AM · Aug 31, 2021 · Twitter Web App

Aaron Kheriaty, MD
@akheriaty

The feature common to all totalitarianisms is not concentration camps or jackbooted soldiers using force--though the deployment of coercion over consent is always a characteristic feature.

The most central feature is the silencing, the forbidding, of certain questions.

11:11 PM · Aug 28, 2021 · Twitter Web App

Dr Clare Craig (not one of her impersonators)
@ClareCraigPath

Don't forget that we are only weeks away from the sacking of the carers we clapped for, who care for the most vulnerable, many of whom have had COVID.

I'm ashamed at the decisions being taken.

11:36 AM · Aug 21, 2021 · Twitter Web App

Wes Pegden
@WesPegden

The greatest "scientific" loss of the pandemic will surely be that so many scientists have forgotten what good scientific evidence looks like.

10:09 AM · Aug 6, 2021 · Twitter Web App

Prof
@covidtweets

"I am fully vaccinated, I always wear a mask (or double), and I got COVID. [Insert some info here about how bad it is]. Get vaccinated and wear a mask so we end this pandemic."

When public heath isn't honest, what you get is people who don't see the issue in this argument.

1:25 AM · Aug 8, 2021 · Twitter for Android

TEAM REALITY

BOUTROS 木
@boutros555

A picture of the COVID-19 pandemic, San Francisco, California, August 2021:

It's been 18 months. I'm still alive. I'm vaxxed. Everyone I know is vaxxed. I didn't get sick. I don't know anyone who got sick.

I go to a restaurant. I have to prove I'm vaxxed.

1/

7:44 PM · Aug 23, 2021 · Twitter Web App

I go from the host stand to my table wearing a mask. The room is full of vaxxed people. They're sitting down, so they're not wearing masks. I sit down so it's ok to take off the mask. Have to pee. I get up to go to the bathroom. I have to put on the mask. ++ I return to the table. It's ok to take off the mask again. The server comes by to take my order. If he's not vaxxed, he gets fired--despite a labor shortage. Thus, he is vaxxed. Everyone is seated at the moment, so he is the only person in the room wearing a mask. ++ Two hours later, it's time to go. I have to wear a mask to the door. I leave the restaurant. I take off the mask. I pass a couple outdoors waiting to get into the restaurant. They are vaxxed, but they are masked. I chit-chat with friends for a bit & then say good night. ++ I walk by the restaurant window & see the couple from earlier sitting at their table. They are now unmasked. We look like crazy people. We. Look. Like. Crazy. People.

Don Wolt
@tlowdon

Heckuva job, American Academy of Pediatrics.

AAP, Aug 5: The Delta variant presents pressing risks to kids, so we urge the FDA to accelerate vaxx approval for ages 5-11.

AAP, Aug 26: There's no evidence that Delta is causing more severe disease in kids than previous strains.

10:50 AM · Aug 27, 2021 · Twitter Web App

Victoria Fox
@drvictoriafox

So we seem to have a new class of people who are "hesitant" about the protection provided by vaccines, yet have absolute faith in the power of masks. I speak to MANY parents and students with this view. How f**ked up does your PH messaging have to be to create that outcome?

10:57 AM · Sep 9, 2021 · Twitter for iPhone

Nick Foy
@TheNickFoy

My son woke up with a scratchy throat today and I'm really angry about it because not many people are wearing masks and nobody is wearing well-fitting N95s people are so selfish and it went away after breakfast and he's fine now but do you know what we've been through 🪦 (1/236)

3:28 PM · Aug 8, 2021 · Twitter for Android

Don Wolt
@tlowdon

CDC DIRECTOR: All pregnant birthing people should be vaxxed.

PFIZER (on 7/28): Um...OK... Yeah, we're not quite done yet with clinical safety testing in expectant mothers.

9:03 PM · Aug 11, 2021 · Twitter for Android

Brumby
@the_brumby

Your child must wear a mask that doesn't work to stop a virus less serious to them than historical respiratory risks until they can get a vaccine they dont need that fails to prevent transmission to teachers/parents who have already been vaccinated anyway

Is that where we're at?

2:43 PM · Aug 12, 2021 · Twitter Web App

David M
@ComradeDoom1

I assume ER doctors regularly treat patients who have done far more heinous things than decline a vaccine.

11:34 AM · Aug 20, 2021 · Twitter Web App

TEAM REALITY

Brumby
@the_brumby

Sweden's 7-day average COVID deaths have been at ZERO for about a month now.

I feel like its only a matter of time before the very existence of a place called Sweden is scrubbed from the internet.

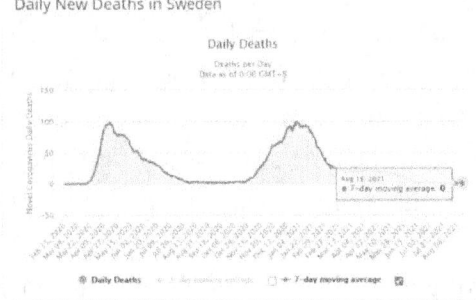

2:21 PM · Aug 17, 2021 · Twitter Web App

John Ziegler
@Zigmanfreud

On day when @AlexBerenson gets suspended from Twitter, highly-vaccinated Israel (the country where Alex went dramatically against the grain), set new all-time record for average news cases.

They're now averaging 5.5 times as many new cases & over 2X as many deaths as a year ago.

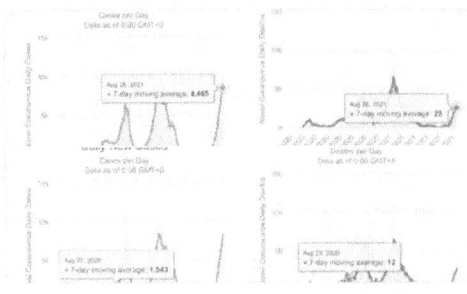

10:40 PM · Aug 28, 2021 · Twitter for iPhone

Phil Kerpen
@kerpen

Canadian mannequin study finds mask filtration efficiencies of:

Cloth: 10%
Blue disposable: 12%
Fitted KN95: 46%
Fitted R95: 60%
KN95 with 3mm gaps: 3%

Even modest ventilation (2 air changes per hour) outperforms best mask.

4:48 PM · Aug 22, 2021 · Twitter Web App

Stefan Baral
@sdbaral

#ZeroCovid will never happen.

1) Vaccines prevent severity, but less so infection.
2) Lots (and lots) of animal reservoirs.
3) Asymptomatic infection (especially with vaccines).

No one likes #COVID19, but it is time to accept that this virus is now part of this world.

Forever.

3:04 PM · Aug 8, 2021 · Twitter Web App

Tracy Hoeg, MD, PhD
@TracyBethHoeg

A telling figure looking at myocarditis rates at 40 hospitals in the Western US w/ recent ⬆ rates correlated w/vaccination. An important thing I would say is COVID itself does not appear to be correlated with an uptick.

I'm NOT anti-vax & am vaccinated.

jamanetwork.com/journals/jama/...

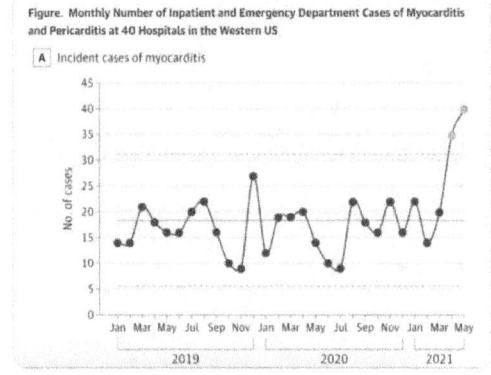

4:30 PM · Aug 5, 2021 · Twitter Web App

Phil Kerpen
@kerpen

The 2009-10 swine flu was five times deadlier to kids than COVID.

... But it didn't hurt boomers, so it didn't stoke an irrational fear/panic that disrupted kids' lives indefinitely.

Pediatric (Age 0-17) Infections and Deaths, Central Estimates

	Infections	Deaths	Infection Fatality Ratio
Swine Flu (April 12, 2009–April 10, 2010)	19,501,004	1,282	0.0066%
COVID (February 2020–May 2021)	26,838,244	332	0.0012%

Sources
https://academic.oup.com/cid/article/52/suppl_1/S75/499147
https://www.cdc.gov/coronavirus/2019-ncov/cases-updates/burden.html

4:09 PM · Aug 7, 2021 · Twitter Web App

TEAM REALITY

Phil Holloway™ @PhilHollowayEsq

Ah. Those were the good old days. All the way in the way back era of 92 days ago.

> **Rochelle Walensky, MD, MPH** @CDCDirector · May 17, 2021
> The science is clear:
>
> If you are vaccinated against #COVID19, you are safe. The vaccines work. You can take off your mask & are not at risk of severe disease or hospitalization.
>
> If you are not vaccinated, you are not safe. Please get vaccinated or continue to wear a mask. twitter.com/nytimes/status...

11:37 AM · Aug 19, 2021 · Twitter Web App

Eric @The_OtherET

It's absolutely insane to me that despite overwhelming evidence showing COVID is spread through aerosols, NOT large droplets, that health organizations are still focusing on masks and the 6 foot rule which are basically useless for aerosolized pathogens

7:26 PM · Aug 15, 2021 · Twitter for iPhone

Eli Klein @TheEliKlein

Enormous numbers of unvaccinated hospital staff are being fired & college students are being disenrolled. We're going to lose unvaccinated military, firefighters, teachers & many more

~50% could be avoided if the CDC followed the science & accepted natural immunity! A disgrace

8:04 AM · Aug 22, 2021 from Manhattan, NY · Twitter for iPhone

---

AJ Kay @AJKayWriter

Repeated, compulsory testing of only unvaccinated individuals is punitive harassment.

Nothing more.

Since both vaxxed & unvaxxed can carry & transmit, singling one out has zero utility.

9:23 PM · Aug 10, 2021 · Twitter for iPhone

Woke Zombie @AWokeZombie

Something to keep in mind: None of these "experts" ever said "Get exercise, eat right, make sure you have proper Vitamin supplements".

Why is that?

1:44 PM · Aug 29, 2021 · Twitter Web App

Craig @TheLawyerCraig

Put aside US politics for a moment: What other countries not only mask 2- to 6-year-olds, but do so with such unwavering certitude that the relative material benefits of all-day cloth masking of this cohort "so obviously" outweighs all drawbacks that no discussion is even needed?

5:02 PM · Aug 12, 2021 · Twitter Web App

Vanessa @vial42

In the past few weeks, I've read babies do not need to see faces to thrive, schools didn't exist a century ago so why bother during a pandemic, all ICU beds for kids are full so your kid will have to wait for another one to die first to MASKS save all lives. When will this end?

4:16 PM · Aug 21, 2021 · Twitter Web App

33 Retweets 3 Quote Tweets 390 Likes

Vinay Prasad, MD MPH @VPrasadMDMPH

Right now America is intoxicated on masking 2 year olds outside and inside ('cept when they nap - virus naps too)

Soon we will be sober and look at what have we done.

It won't age well, I promise.

4:44 PM · Aug 22, 2021 · Twitter Web App

Show Me The Data @txsalth2o

We were right about age risk
We were right about beaches
We were right about schools
We were right about plexiglass
We were right about fomites

We WERE RIGHT ABOUT NATURAL IMMUNITY

man it's going to be weird when you realize we were right about masks

9:16 AM · Aug 27, 2021 · Twitter Web App

Matt, Pre-School Diploma @statomattic

It is possible to believe...

COVID is real AND lockdowns and masks are ineffective and wrong.

The vaccines work AND so does natural immunity AND most kids don't need a vaccine.

We must limit death WITHOUT limiting life.

Science is wonderful AND "The Science™" is not science.

10:21 AM · Aug 5, 2021 · Twitter for Android

TEAM REALITY

Angry Cardiologist
@AngryCardio

Too many of my colleagues, facing the reality of vaccine-associated myocarditis, are either burying their heads in the sand or throwing up their hands.

In an effort to promote vaccination, they are making what I believe are misguided actions.

12:12 AM · Aug 21, 2021 · Twitter for iPhone

I believe vaccination for COVID is our best way back to a normal life. And we need to be honest with people about what we know. We also need to acknowledge that there is a great deal we can do to minimize the harms of our interventions. ++ Regarding what we know—we need to be honest about who is affected, and how frequently. We shouldn't try to make marginally valid comparisons to COVID. The folks you want to persuade won't believe you anyway. Most people think differently about an active interventions & disease. ++ Regarding what we can do, here are a few things: 1) Dose those at myocarditis risk differently, and/or with different vaccines. 2) Advocate for alternative vaccination / natural immunity to be accepted in schools, businesses, etc. ++ 3) Advocate for compensation for affected patients & families. Vaccine-associated myocarditis rarely kills, but does impart a financial burden. These are things every physician can do. No special skills or equipment needed.

[*CDC, six months after this thread: "An 8-week interval may be optimal for some people ages 12 years and older, especially for males ages 12 to 39 years." __Prof*]

Martin Kulldorff
@MartinKulldorff

For thousands of years, disease pathogens have spread from person to person. Never before have carriers been blamed for infecting the next sick person. That is a very dangerous ideology.

11:04 AM · Aug 24, 2021 · Twitter Web App

This was a remarkable tweet. Dr. Kulldorff undoubtedly knew this not to be true, but the tweet unleashed a mob of "experts" giving examples of past atrocities to justify the current one. _Prof

Prof
@covidtweets

I can't wear a mask in grocery stores or when I am walking to my table in restaurants because I have this, medical condition called critical thinking. It is terrible...

4:21 PM · Aug 20, 2021 · Twitter for Android

Prof Francois Balloux @
@BallouxFrancois

Anecdotes about someone vaccinated getting infected by COVID, and feeling poorly for a couple of days should be presented as 'good news' stories, and a triumph of science, not as portents of the Apocalypse.

8:46 AM · Aug 22, 2021 · Twitter Web App

Sarah Beth Burwick
@sarahbeth345

The biggest issue I have with masks right now is how much of a false sense of security they are giving high-risk people. The "masks save lives" narrative makes people think they are invincible as long as they have a piece of cloth on their faces.

8:42 PM · Aug 28, 2021 · Twitter for iPhone

District AI
@districtai

I only want one question answered:

Where is the data or RCT that shows masking works? The CDC isn't providing it.

The answer is they don't have it.

9:48 PM · Aug 18, 2021 · Twitter Web App

TEAM REALITY

ClownBasket
@ClownBasket

So kids are in masks in school because they can't get vaxed.

Yet vaxes don't stop infection or spread—they reduce symptoms.

Yet kids naturally have reduced symptoms & spread without vaxes.

So what are we doing?

3:09 PM · Aug 20, 2021 · Twitter for iPhone

Kyle Lamb @kylamb8 · Aug 26, 2021
Apparently the Sturgis rally has been a multi-state superspreader off and on for 18 months...

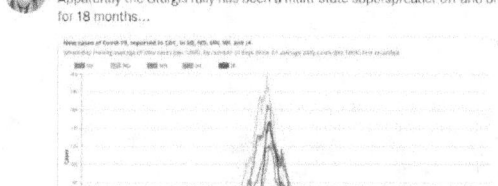

Infectious Disease Ethics
@ID_ethics

Falsehoods that have undermined rational #covid19 policy

-Immunity denial
-Age-severity curve denial
-Seasonality denial
-Outdoor risk = indoor
-Asymptomatic risk = symptomatic
-Impossibility of shielding the vulnerable
-No public health harm from lockdown

Any more?

9:44 PM · Aug 31, 2021 · Twitter Web App

Anish Koka
@anish_koka

I learned in 2021 that society exists to keep hospitals empty.

9:05 AM · Aug 2, 2021 · Twitter for iPhone

Sarah Beth Burwick
@sarahbeth345

I just read a book to my 2yo. She sat in my lap, and looked up at me during funny parts to see my face and confirm I was smiling.

If you contend that masks do not affect a toddler's social & emotional learning, you have lost all credibility. I don't trust a word you say.

10:20 PM · Aug 24, 2021 · Twitter for iPhone

Eli Klein
@TheEliKlein

My accurate quote from Science, one of the leading scientific journals, was marked misleading by Twitter, despite also being tweeted by 1000s & covered by many leading experts. I didn't write against recommending vaccination, as the misleading label insinuates. This is defamation

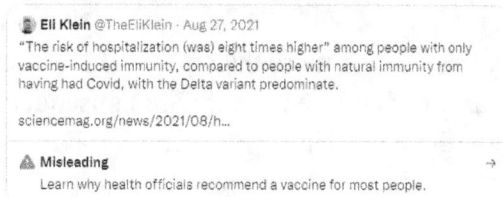

⚠ Misleading
Learn why health officials recommend a vaccine for most people.

5:22 PM · Aug 27, 2021 from Manhattan, NY · Twitter for iPhone

---

Eric
@The_OtherET

I'm seeing a growing sentiment that people should be prioritized or denied care based on vaccination status and all I can say is that this is a dangerous and evil line of thinking. There are plenty of other preventable diseases that are treated despite patients' "wrong" choices.

10:44 PM · Aug 16, 2021 · Twitter for iPhone

Dr Clare Craig (not one of her impersonators)
@ClareCraigPath

Eating disorder referrals have soared thanks to government policy.

Anorexia Nervosa patients have a 5 x increased mortality and for Bullemia patients it is doubled.

When do these mostly young people start to matter?

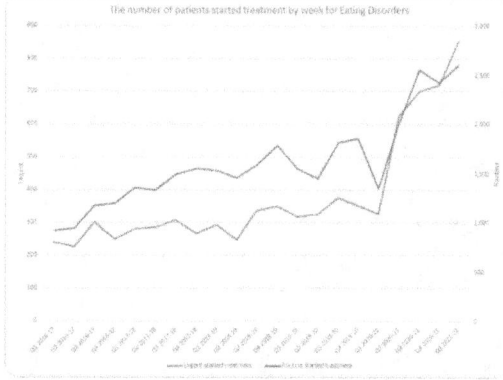

3:16 AM · Aug 19, 2021 · Twitter Web App

Bethany S. Mandel ✓
@bethanyshondark

This is not about homeschooling. This is not about winning more converts to my side. This is simply a plea from one parent to another: If your kid's school resembles a prison, PULL THEM OUT. It is early in the year. There is time.

8:08 AM · Aug 26, 2021 · Twitter for iPhone

Your kids get one childhood. One chance at their development. If they spend the vast majority of their waking hours in these buildings, you need to ask what it will do to them and if it's worth it. It is up to us as parents to ensure their souls aren't damaged by this. ++ It's more than masks. It's plastic partitions. It's fear. It's not being allowed near other children. There are consequences to all of this and innocent children will bear them. ++ There are a million homeschool resources out there. I understand it's scary and overwhelming but the damage COVID lunacy will do to kids is scarier.

Sarah Beth Burwick
@sarahbeth345

How the left is losing me, no. 8

Everything is an absolutist purity test. If you question masks, then you are a covid-denying anti vaxxer. If you subscribe to a right-wing voice on any point, then you are a racist homophobic misogynist. No nuance.

#howtheleftislosingme

12:08 AM · Aug 26, 2021 · Twitter for iPhone

Eric
@The_OtherET

The CDC and other public health organizations have had over a year and a half to perform randomized controlled trials on masks in schools, but haven't.

One has to start wondering if they are so averse to such a gold-standard method because of what the findings could suggest.

3:04 PM · Aug 10, 2021 · Twitter Web App

Anish Koka
@anish_koka

Think this through.

Who does a city wide vax mandate hurt more ? The high $$$ restaurant in the nice neighborhood where 100% are vaxxed .. or the little deli in a rougher part of town where the majority of customers may be unvaxxed?

10:26 AM · Aug 4, 2021 · Twitter for iPhone

Jenin Younes
@Leftylockdowns1

Text from a vaccinated friend: "I will never show my ID and vax card, no matter what i miss out on." We need more people like you, friend who's name I promised not to use here.

9:34 PM · Aug 14, 2021 · Twitter for iPhone

Emma Woodhouse
@EWoodhouse7

Our Governor is so logical that tonight, our daughter could sit maskless at a restaurant table that's ~500 meters away from the school in which she's *currently* enrolled, after a day of not being able to do the same at her classroom desk. #TheScience

10:49 PM · Aug 27, 2021 · Twitter for iPhone

Emily Burns ✓ DMs welcome #TeamReality
@Emily_Burns_V

1/
Most states are seeing significant increases in cases over last summer--even testing-adjusted. The YoY increases are LARGER in the 10 most-vaxed vs. 10 least-vaxed states—with VT (most-vaxed) one of the highest.

h/t @kerpen for building tool.
docs.google.com/spreadsheets/d...

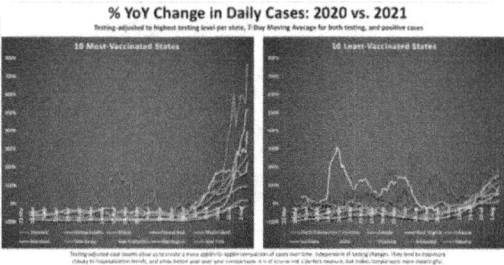

7:19 PM · Aug 16, 2021 · Twitter Web App

Vinay Prasad, MD MPH 🎤📺 ✓
@VPrasadMDMPH

There will not be a level of vaccination (with or without boosters) that will bring the risk of acquiring sars-cov-2 to zero;
The sooner we accept that, the sooner we can have sensible health policy

7:08 PM · Aug 24, 2021 · Twitter Web App

Emily Burns 😊 DMs welcome #TeamReality
@Emily_Burns_V

1/
Whether it's MAGA Republicans, or inner city minorities, vax hesitancy seems to be rooted in distrust of those in power. If the health of these disenchanted groups is the goal, mandates seem like the wrong tactic. Let's see how they fare w/other vaxes.
mass.gov/doc/weekly-cov...

ation Rates in Boston, MA by Race (8/30/		
Race/Ethnicity	Individuals with at least one dose per capita	Propo... po...
ian	70%	
ack	43%	
spanic	44%	
hite	61%	

4:27 PM · Aug 30, 2021 · Twitter Web App

15 of 50 U.S. states allow for philosophical and religious exemptions to school vaccine mandates. The states w/no exemptions at all actually see the LOWEST vax rates for full vaccination of the 7-vax child vaccine series by 35 mos. ++ Interestingly, states of all types get well above 90% with the MMR vax by the time kindergarten rolls around—these are after all, deadly diseases. But these data exclude children pushed out of schools by states that do not allow religious or philosophical exemptions. ++ Let's take a look then, at the 2019 measles outbreak, and see if states without exemptions fared better, or worse than those with exemptions. As you can see, they fared worse—quite a bit worse, actually. ++ I believe this is because mandates without exemption marginalize people. Indeed it is their explicit goal—one I believe is both unwise and unethical. Marginalized people are forced to become a drag on society. At worst they become radicalized. ++ What happens to someone who loses their job and can't get another due to vax mandates? They go on welfare, and Medicaid—we are forcing them into poverty. Why would we do that for a choice that impacts only them? ++ I think there's a desire to punish some gap-toothed redneck down South with these policies. But throughout the country, the people these policies will harm most are minorities, particularly urban minorities. ++ This happens with so many of our policies—in pursuit of some perceived utopia, we implement policies that harm the most at-risk within our society. COVID policy has been a masterclass in this. ++ There might have been an argument for this when we (erroneously, I believe) thought the vaccines conferred sterilizing immunity. Now I think it is just a cop-out on the part of the CDC. ++ I find stories like this both nauseating, and heart-breaking. Nauseating to watch people who claim to work in the service of health glorying in someone's death. Heart-breaking because I believe that the vaccines likely would have helped him. ++ And there is one culprit—the CDC. Mandates are easy, conveying relative risk is hard. But that is what the CDC needs to do. Until they breakdown the risk to people by co-morbidity, we will continue to have people die unnecessarily. ++ Again, the CDC and the media seem to be happy when these people fall into their "enemies" camp. But far more often, per national vaccination trends, they will fall into their "friends" camp. ++ Many of us have been calling for this since LAST May, have even tried to piece it together ourselves. The CDC has 20K employees. Assign each one 30 deaths, and figure out the co-morbidities. Then do outreach in a targeted fashion. ++ But this is exactly what public health DOESN'T want to do. It's boring, and it doesn't get the media attention. But it gets results--something it's fairly clear the CDC is OK with foregoing. ++ Interestingly, you can see how different an approach you get--one based on convincing, rather than coercing--from Dr. Osterholm. MN has a philosophical exemption, so he KNOWS that he has to convince people first. This should be the way of public health.
https://pbs.org/wnet/amanpour-and-company/video/do-masks-provide-much-protection-we-think-bglhwy/

TEAM REALITY

 **Jay Bhattacharya**
@DrJBhattacharya

Mortality from #COVID19 differs more than a thousand-fold between the old and young. Focused protection is the compassionate approach that balances COVID risks and collateral damage to public health.

gbdeclaration.org
Great Barrington Declaration and Petition
As infectious disease epidemiologists and public health scientists we have grave concerns about the damaging physical and mental health impacts of the ...

5:28 PM · Aug 23, 2021 · Twitter Web App

 **Zac Bissonnette**
@ZacBissonnette

Has anyone ever been suspended from Twitter for exaggerating the risk covid poses to kids?

8:40 PM · Aug 28, 2021 · Twitter for iPhone

 **Jennifer Sey** ✓
@JenniferSey

Weekly reminder, in Denver where I live, adults can go to large scale sporting events, night clubs, the mall, bars & strip clubs unmasked. But 2 year olds learning to speak need to mask for 8 hours a day in pre school and daycare. Does anyone really think this makes sense?

12:43 PM · Aug 27, 2021 · Twitter for iPhone

 **elizabeth bennett**
@ebennett74

Gov is mandating masks for all children in school. My special needs kid is hearing impaired and often hard to understand. Asked him if he wanted a medical exemption; he said no—he's starting a new high school and doesn't want to stand out. Reason 1000 why mask mandates suck.

7:39 PM · Aug 19, 2021 · Twitter for iPhone

 **Eric**
@The_OtherET

"This new COVID variant is SO transmissible guys, that's why it's so important we do all the shit that didn't work the last time"

11:37 PM · Aug 8, 2021 · Twitter for iPhone

 **Jenin Younes**
@Leftylockdowns1

I literally cannot believe around half this country sees no problem with forcing others to stay home, close their businesses, get a vaccine for a disease that in many cases presents virtually no risk to them, and wear a piece of cloth over their faces.

12:07 PM · Aug 25, 2021 · Twitter Web App

 **Erich Hartmann**
@erichhartmann

Wow.
The AAP is actively "deleting" entire sections from their website re early childhood development & the importance of facial cues for learning. They are memory holing decades of known & accepted medicine, all bc they have embraced forced masking of our nations children.
Wow.

4:41 PM · Aug 18, 2021 · Twitter Web App

TEAM REALITY

SEPTEMBER 2021

Jenin Younes
@Leftylockdowns1

Seen in a comment: "the protected need to be protected from the unprotected by forcing the unprotected to use the protection that didn't protect the protected." Anonymous internet user, you are a genius

5:11 PM · Sep 21, 2021 · Twitter for iPhone

TEAM REALITY

Erich Hartmann @erichhartmann

Dr. Danny "No Control Group Needed" Benjamin gives parents advice on how to mask toddlers. Listen up:

5:31 PM · Sep 25, 2021 · Twitter Web App

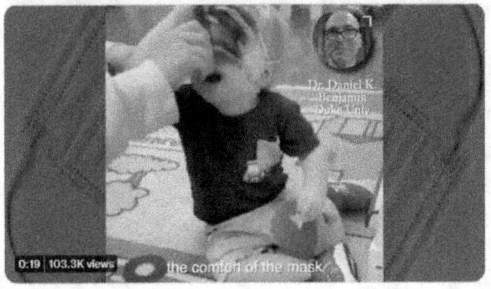

5:31 PM · Sep 25, 2021 · Twitter Web App

Sarah Beth Burwick @sarahbeth345

"My kid doesn't mind wearing a mask" is the most twisted mom flex ever. I don't WANT my kids to be content with their faces covered. I want my kids to hate it, and I hate it too, because we are normal humans who desire social connections and fresh air.

12:44 AM · Sep 3, 2021 · Twitter for iPhone

Virál Myãlgia MD, PhD @contrarian4data

As @JoeBiden asks Americans to blame the unvaccinated for the spread of a respiratory virus, the Gibraltarians hunted down the <1% who caused a "summer wave" that was larger than in '20.

That <1% may cause more problems this winter.

If this makes sense to you, you are <insert>.

5:37 PM · Sep 9, 2021 · Twitter Web App

Anish Koka @anish_koka

Imagine if the @CDCDirector framed the conversation on vaccines this way.

Bizarro world where well spoken NBA players have more nuanced conversations on vaccine policy than govt. public health professionals.

From **RK**

8:57 AM · Sep 28, 2021 · Twitter for iPhone

Bachman @ElonBachman

A local blood drive is now requiring COVID vaccination

So first they sidelined healthy nurses and docs, creating staff shortages

Now blood

These hysterics are really going to will a medical crisis into existence where none existed

10:52 AM · Sep 11, 2021 · Twitter Web App

M_P @Reroot_Flyover

The CDC is arguably the biggest purveyor of Covid misinformation out there. That is a dire problem.

6:48 PM · Sep 12, 2021 · Twitter Web App

Tracy Høeg, MD, PhD @TracyBethHoeg

UK and Norway are now making the same recommendation for 12-15 year olds - offer children 1 dose of Pfizer.

> **BBC Breaking News** @BBCBreaking · Sep 13, 2021
> All 12 to 15-year-olds should be offered one dose of Pfizer coronavirus vaccine, UK's chief medical officers recommend bbc.in/3E5yrxs

10:20 AM · Sep 13, 2021 · Twitter Web App

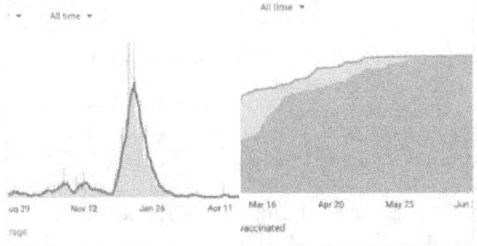

TEAM REALITY

ClownBasket 😊
@ClownBasket

Replying to @drterryrmcd

We are going with the "restaurant rules" for any guests in our house. Masks on while walking from point A to point B. But once seated their masks can come off.

Nothing beats the protection of a mask. Nothing.

10:23 PM · Sep 29, 2021 · Twitter for iPhone

Infectious Disease Ethics
@ID_ethics

Since when did public health move from:

"protecting the vulnerable"

to

"protecting the vaccinated" ?

> **Kamala Harris** @KamalaHarris · Sep 12, 2021
> United States government official
> By vaccinating the unvaccinated, increasing our testing and masking, and protecting the vaccinated, we can end this pandemic. That's exactly what we are committed to doing.

11:39 PM · Sep 12, 2021 · Twitter Web App

Brumby
@the_brumby

Whatever ridiculous 6-pronged bullshit Biden trots out today, it will have an equal chance of altering the course of earth's magnetic field as it does the course of COVID in the US.

Get a vax if you want, leave my kids alone, & let all move on w life however they're comfortable.

1:58 PM · Sep 9, 2021 · Twitter Web App

Stinson Norwood
@snorman1776

The naturally immune should be paid reparations for every week they've been immune and continued to have their lives constrained.

12:44 PM · Sep 16, 2021 · Twitter Web App

Martin Kulldorff
@MartinKulldorff

Friend: "Be safe"
Me: "Enjoy life"

6:06 PM · Sep 23, 2021 · Twitter Web App

Sarah Beth Burwick
@sarahbeth345

2020: hate has no home here

2021: unvaccinated people should choke on a dick and die alone

#HowTheLeftIsLosingMe

7:18 PM · Sep 30, 2021 · Twitter for iPhone

Show Me The Data
@txsaith2o

Yesterday my daughter had 14 kids yelling at her to put on a mask, she refused.

She discussed why masks were ineffective & unnecessary, their response "it doesn't matter, that's the rule"

It doesn't matter, that's the rule.

9:00 AM · Sep 17, 2021 · Twitter Web App

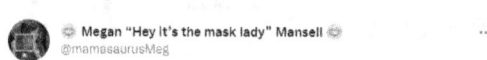
Megan "Hey it's the mask lady" Mansell
@mamasaurusMeg

Why are there quarantine measures for something with a 99.8% survivability rate and near-0 IFR for children, yet you can return to school from RSV, strep, or flu as soon as you feel better, which are far deadlier for children (and you'll never be notified of exposure)?!?!?

2:54 PM · Sep 14, 2021 · Twitter for iPhone

John Ziegler ✓
@Zigmanfreud

I REALLY wish people, especially in the media, understood that "scientific/expert consensus," especially in highly-charged topics, is often born NOT of overwhelming facts/evidence, but rather due to HUGE incentives created for everyone to come to & not contradict that conclusion.

11:18 AM · Sep 24, 2021 · Twitter for iPhone

Wes Pegden
@WesPegden

There are a lot of reasonable things to put in a 6 prong plan. But one thing that should definitely be there is doing the hard work to increase vaccination rates among the older and most vulnerable populations.

Even in August data, the median COVID death age is still around 70.

7:26 PM · Sep 8, 2021 · Twitter for Android

Vinay Prasad, MD MPH
@VPrasadMDMPH

The CDC director, a political appointee, siding with the President, & his pre-ordained plan, against the advice of the ACIP advisors, with 0 RCTs measuring clinical outcomes & no idea of adverse events is not "doing one's job" it is politics > science, precisely what we feared 🪦

11:37 AM · Sep 24, 2021 · Twitter Web App

The process that led up to Pfizer booster is someday going to be seen as what many dreaded during the Trump administration. The White House decided during the week of Sept 20, boosters will be made available, and all gears of drug regulation made it happen ++ Consider the 24 year old male health care aid living in Seattle who received 2 doses of Pfizer vax, and is now granted a path to take dose 3. Can any medical expert answer 2 questions: Do the benefits exceed the harms to this individual? Is he a 'safer' employee? ++ The correct answer is that no one knows. That is kinda a huge error of medical regulation. A healthy 24 year old will receive a vaccine product, and we do not data to answer 2 simple qs. How about another e.g. ++ A 25 year old, worried teacher, who had documented sars-cov-2 & recovered, 2 doses of Pfizer and works in LA. This teacher has a path to booster, but is it: In their best interest (benefit/ harm)? Benefit to others in their school? The right ans = no one knows ++ And, It is entirely possible the third dose is a net detriment to them. Walid Gelad is right ++ One purpose of drug and vaccine regulation is to protect people-- avg. worried, concerned, nonchalant, indifferent, invested-- people from making choices that may worsen their health outcomes. this policy undermines that protection ++ If instead all you want is total autonomy to boost-- then why have the FDA at all. Just live in a world where you can get all the boosters you want when you want. That world would be brutish and short ++ As a general rule, the standards we use for very ill people (relapsed cancer) are more lax than very healthy people (very high) In emergencies, the standard can fall. Having said all that, it is highly questionable to claim.. ++ that a 25 year old who recovered from the virus and had 2 doses of pfizer faces an *emergency* such that the regulatory standard of *total speculation* is acceptable. Misusing expedited & lower evidence pathways to push products is a common failure at FDA ++ Politicians cannot drive these decisions. Politicians want fewer cases -- end of story-- famously one asked that a Boat not dock at long beach to avoid the case count rising. Not testing sick people is another way to depress cases artificially. But also... ++ Boosting healthy people may reduce the rate some people test + on PCR, while not changing rates of severe illness, a tiny fraction may have serious AEs that offset any gains A politician is not invested in the health of people, but the political implications ++ Finally, vaccine safety is a long game. Politicians have short term horizons, but any error here will destroy us.. Consider this... If it is shown that healthy young (20 year old) health care workers getting booster results in a net health decrement to them (I'm not saying it does, IDK, but we can't exclude it), then what have you done to health care worker acceptance of vaccination... ++ You make any mistake here and you salt the earth. ++ As someone who has studied regulatory science for more than a decade, I would not have made this gamble. Politicians think differently. Bad process. ++ Worse IMO is how many who would criticize this if trump did it, but when Biden does... dark days for science ahead.

AJ Kay
@AJKayWriter

Well, they have a new customer now.

Walked in after seeing this sign and spent more money that I needed to — probably more than I should've.

Because I'm pro-business, pro-human, and motivated to support the brave people who support me.

9:14 PM · Sep 21, 2021 · Twitter for iPhone

Eric
@The_OtherET

This pretty much summarizes my thoughts on teachers unions, the ACLU and the CDC over the past year and a half

8:43 PM · Sep 3, 2021 · Twitter for iPhone

Prof Francois Balloux
@BallouxFrancois

For the next pandemic, I suggest that everyone, the general public, politicians, journalists and scientists get a crash course in probability, statistics and experimental design at the start. It might not save that many lives, but it would at least elevate the debate.

8:24 PM · Sep 1, 2021 · Twitter Web App

BOUTROS 木
@boutros565

Six prongs. Six.

And none of them target increasing vaccination among the most at-risk age group (65+).

4:37 PM · Sep 9, 2021 · Twitter Web App

Matt, Pre-School Diploma
@etatomattic

Full NFL stadiums today from coast to coast. As it should be. And, in two weeks, national COVID cases will be lower than they are today.

The summer Delta wave is already waning. Because seasonality, not human activity, drives these patterns.

Rejoice. Reality shall prevail.

9:52 AM · Sep 12, 2021 · Twitter for Android

Jay Bhattacharya
@DrJBhattacharya

The history of medicine is filled with examples of things known true until proven false by randomized trials.

19 months into the pandemic, we have no RCTs on masking children & "experts" who swear masking toddlers is essential for COVID safety (except while napping).

4:01 PM · Sep 27, 2021 · Twitter Web App

TEAM REALITY

Kyle Lamb
@kylamb8

Scott Gottlieb acknowledged just now on Fox News Primetime that the 6 feet distancing was just winging it and based on science that 'wasn't really established.' More or less credited Florida for doing 3 feet early on, rather than 6.

History scores another win in Florida's favor

7:12 PM · Sep 27, 2021 · Twitter Web App

Prof
@covidtweets

Fauci, Jan 2021: If one mask is good, it is common sense that two would be better.

Fauci, Sep 2021: If two doses are good, three will be probably better.

This is how we science...

> There was good reason to believe that a third dose "will actually be durable, and if it is durable, then you're going to have very likely a three-dose regimen being the routine regimen," Fauci, the director of the National Institute of Allergy and Infectious Diseases, said at a briefing Thursday.

7:51 AM · Sep 3, 2021 · Twitter for Android

Virál Myãlgía MD, PhD
@contrarian4data

As @JoeBiden asks Americans to blame the unvaccinated for the spread of a respiratory virus, the Gibraltarians hunted down the <1% who caused a "summer wave" that was larger than in '20.

That <1% may cause more problems this winter.

If this makes sense to you, you are <insert>.

5:37 PM · Sep 9, 2021 · Twitter Web App

Sarah Beth Burwick
@sarahbeth346

Is anyone in the US studying or reporting on how covid is playing out in African countries?

I just looked up stats in Ghana, for example, and am baffled. Population of 31M. Only 1100 total covid deaths. Less than 3% of population fully vaccinated.

Huh?

9:49 PM · Sep 21, 2021 · Twitter for iPhone

Prof
@covidtweets

There are millions of people who still believe that once a certain percentage of the population is vaccinated, Covid will end.

That's the reason for the hostility against those who choose not to get vaccinated.

It will be a hard awakening.

11:37 PM · Sep 22, 2021 · Twitter for iPhone

Ann Bauer
@annbauerwriter

For the record, I'm not against the vaccine. I took the vaccine. I'm glad I did.

What I'm against is a president's appearing on television and telling one group of Americans to fear and shun another group of Americans.

That's not American. It's what totalitarian dictators do.

8:38 PM · Sep 11, 2021 · Twitter Web App

Hold2
@Hold2LLC

Gents, look at this new model.

They acknowledge seasonality - and the fact it showed clearly last Fall - but did not input into the model?

Bold move predicting the opposite of historical respiratory illness patterns.

npr.org/sections/healt...

@districtal @covidtweets

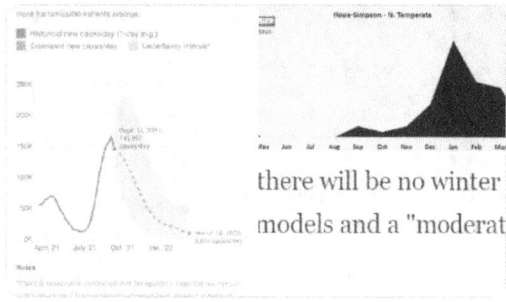

11:28 AM · Sep 22, 2021 · Twitter Web App

John Ziegler ✓
@Zigmanfreud

If not for the mostly packed and almost totally unmasked football stadiums this season, there would be NO hope left for what used to be the greatest nation humans ever created...

3:40 PM · Sep 18, 2021 · Twitter for iPhone

TEAM REALITY

Sarah Beth Burwick
@sarahbeth345

A fully vaccinated HS teacher, who wears a mask and face shield to teach her fully vaccinated students, just told me she hates when students come close to ask her a question.

3:49 PM · Sep 26, 2021 · Twitter for iPhone

BOUTROS 木
@boutros555

You're eventually going to get infected with SARS-COV-2, the virus that causes COVID.

You need to get over it.

9:01 PM · Sep 2, 2021 · Twitter Web App

Craig
@TheLawyerCraig

15 of the 16 states below the 37th parallel are currently showing a decrease in the #Covid19 hospitalization rate compared to two weeks ago (data: @nytimes). South Carolina is the only one up, by 3%. The US is split about in half for increase vs. decrease.

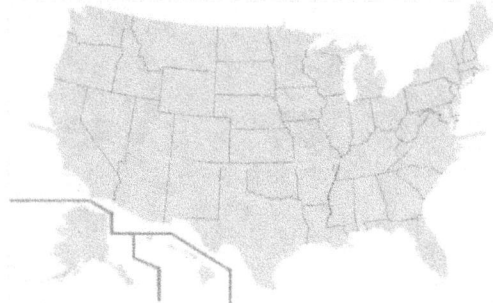

2:22 PM · Sep 20, 2021 · Twitter Web App

Real Developments
@pdubdev

Biden admin pushing dishonorable discharges for troops that refuse vaccinations.

a totally shameful punishment for the men and women who have spent their lives fighting for our freedom

11:08 AM · Sep 23, 2021 · Twitter Web App

Phil Holloway™
@PhilHollowayEsq

iPad school is lethal. What we've done to the kids in order to salve the #COVID19 anxiety of adults is unforgivable: 3 out of 5 teens who died by suicide in Milwaukee County last year cited 'virtual learning' as a stressor

8:40 AM · Sep 6, 2021 · Twitter for iPad

Emma Woodhouse
@EWoodhouse7

My radical plan for schools

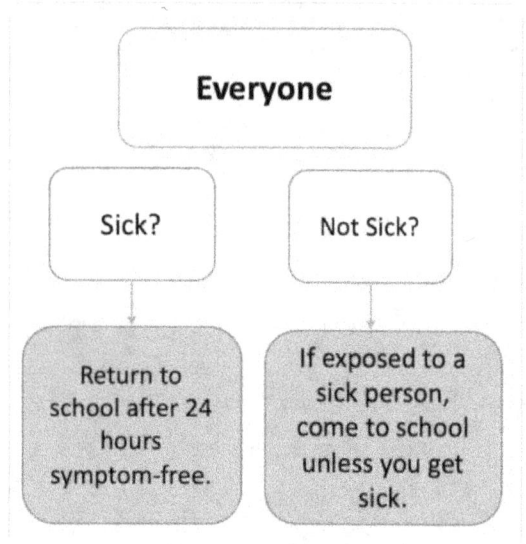

8:54 AM · Sep 18, 2021 · Twitter for iPhone

Justin Hart
@justin_hart

What would we do without our unelected health overlords? Guess what - you're free to kiss again in Oregon!

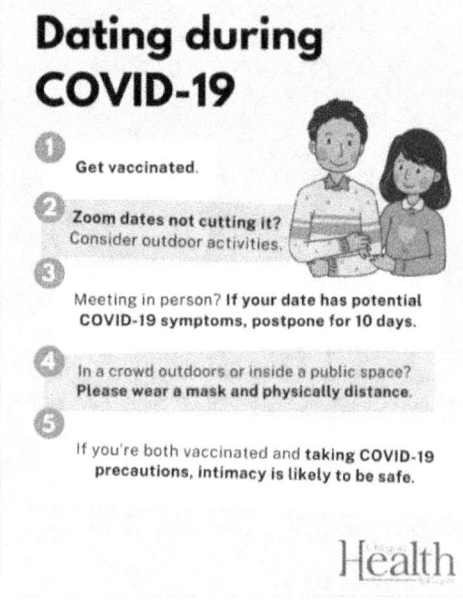

7:36 PM · Sep 23, 2021 · Twitter Web App

TEAM REALITY

Angry Cardiologist
@AngryCardio

Non-COVID vaccinated only

What is the main reason you are not vaccinated?

Vaccine not safe	54.6%
Vaccine not effective	15.4%
Busy, no time to get	1.6%
Other (please comment)	28.4%

20,479 votes · Final results

7:35 AM · Sep 11, 2021 · Twitter for iPhone

Eli Klein
@TheEliKlein

1000s of NY's unvaccinated hospital workers are already out of a job.

This is not proof that "vaccine mandates work."

Don't let those trying to paint this bright picture fool you.

This is a betrayal of heroes who worked through tough Covid times and have natural immunity.

7:18 AM · Sep 29, 2021 from Manhattan, NY · Twitter for iPhone

---

Jay Bhattacharya
@DrJBhattacharya

During COVID, public health has conditioned us to think of others as clean and unclean:
1) careful / reckless
2) masked / unmasked
3) uninfected / infected
4) vaxxed / unvaxxed
5) boosted / unboosted (soon)

We must resist and seek to love our neighbors as ourselves instead.

9:02 PM · Sep 10, 2021 · Twitter Web App

Prof Francois Balloux
@BallouxFrancois

A depressing lesson from the covid pandemic is that pre-existing health inequities have likely been stronger determinants of morbidity and mortality overall than any pandemic mitigation measure.
1/

3:56 AM · Sep 27, 2021 · Twitter Web App

elizabeth bennett
@ebennett74

Every doctor i know today got an email from their medical board threatening to revoke their license if they "spread misinformation" about COVID vaccines. The boards get to define "misinformation" however they choose. Is it "misinformation" to discuss known serious adverse rxns? 😳

6:35 PM · Sep 9, 2021 · Twitter for iPhone

Josh Stevenson
@ifihadastick

THREAD:

Posting Every Randomized Control Trial on Masking kids & Mask mandates in schools: Peer Reviewd Medical and Academic publications only with links!

5:41 PM · Sep 26, 2021 · Twitter for iPhone

[No tweets follow, since there are zero RCTs on masking kids _prof]

Vanessa
@vlal42

Just spoke to a friend in LA- her 5 year old daughter's hands are cracked and bleeding from all the hand sanitizer they apply at school. Can we make the madness stop? The long term psychological ramifications will be catastrophic.

9:38 AM · Sep 16, 2021 · Twitter for iPhone

Eric
@The_OtherET

The rich and famous can't spread COVID. Masking is solely reserved for us peasants, vaccinated or not.

10:54 PM · Sep 19, 2021 · Twitter for iPhone

Gummi Bear
@gummibear737

My message to unvaccinated people

1) I stand with you, whatever your decision is

2) As a vaccinated person I neither judge nor fear you... your vax status doesn't affect me

3) If you're old, obese and/or have co-morbidities, reconsider taking the vaccine...it might save your life

5:28 PM · Sep 9, 2021 · Twitter for iPad

TEAM REALITY

Emma Woodhouse
@EWoodhouse7

Two students, neither sick:

Stu A is vaccinated, Covid-positive, & asymptomatic

Stu B is unvaccinated, Covid-positive, & asymptomatic

Stu A doesn't have to miss any school

Stu B has to miss 10 days

Logical?

10:27 AM · Sep 18, 2021 · Twitter for iPhone

Martin Kulldorff
@MartinKulldorff

Many hospitals demand that employees with prior COVID disease get vaccinated, despite already having stronger longer lasting immunity. If hospitals cannot get the medical evidence right on this one, how can we trust them with other aspects of our health?

> **Martin Kulldorff** @MartinKulldorff · Aug 25, 2021
> In Israel, vaccinated individuals had 27 times higher risk of symptomatic COVID infection compared to those with natural immunity from prior COVID disease [95%CI:13-57, adjusted for time of vaccine/disease]. No COVID deaths in either group.
> medrxiv.org/content/10.110...

10:05 AM · Sep 13, 2021 · Twitter Web App

M_P
@Reroot_Flyover

I'm beginning to worry that Covid is too big to fail.

9:59 PM · Sep 28, 2021 · Twitter Web App

elizabeth bennett
@ebennett74

Come on, college students: push back against these insane restrictions. You are paying a pretty penny to be treated like lepers.

12:26 AM · Sep 22, 2021 · Twitter for iPhone

Erich Hartmann
@erichhartmann

One of the main reasons I speak up, and speak out against this COVID insanity is I know a lot of people who agree... but cannot speak up, because if they did they'd lose their jobs. And that's not okay. None of this is okay.

12:02 PM · Sep 21, 2021 · Twitter Web App

Eric
@The_OtherET

Seeing a lot of fearmongering on major news networks about COVID's danger to kids, so just thought I'd point out that in 2018 the peak pediatric hospitalization rate for flu was 3.4/100k, while for COVID it's less than half at 1.6, yet I don't remember widespread panic in 2018

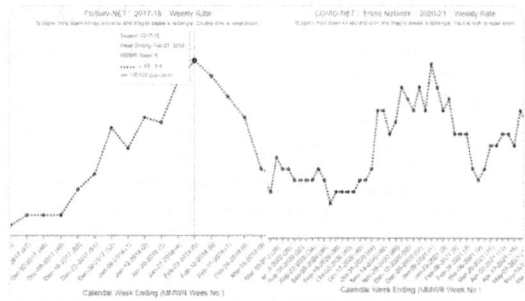

2:27 PM · Sep 20, 2021 · Twitter Web App

Jay Bhattacharya
@DrJBhattacharya

Opposition to discriminatory vaccine mandates is not the same as opposition to vaccines. On the contrary, support for vaccine mandates is an anti-vax position because it breeds distrust and resentment toward public health.

6:21 PM · Sep 4, 2021 · Twitter Web App

Prof Francois Balloux
@BallouxFrancois

A variant a day keeps the doomers in play.

6:51 AM · Sep 1, 2021 · Twitter Web App

Zac Bissonnette
@ZacBissonnette

Frustrating thing:

The reason we have this enormously disruptive testing and quarantining regimes in schools is that it's easier to keep kids out of school and shove stuff in their noses than to get Fauci to admit he exaggerated the risk of schools in covid spread.

3:52 PM · Sep 9, 2021 · Twitter for iPhone

Jennifer Sey
@JenniferSey

I don't get the folks who are too afraid to work in person but just got back from Hawaii or Mexico or Greece or Aruba or wherever else on vacation.

1:40 AM · Sep 8, 2021 · Twitter for iPhone

TEAM REALITY

OCTOBER 2021

BOUTROS 木
@boutros555

Emergencies don't last 19 months.

10:04 AM · Oct 8, 2021 · Twitter Web App

TEAM REALITY

Anish Koka @anish_koka

It has been 19 months since the pandemic started. I'm a doctor. I take care of patients in hospitals every day. The visitation restrictions still in place at hospitals are inhumane, cruel and pointless.

6:12 PM · Oct 21, 2021 · Twitter for iPhone

Eli Klein @TheEliKlein

Los Angeles is requiring healthy teenage boys to take 2 mRNA shots in order to visit malls, museums, etc.

Most of the world recommends against a 2nd mRNA shot for this demographic—no country mandates it.

This Covid vaccine extremism destroys trust public health.

So stupid.

11:10 PM · Oct 6, 2021 from Manhattan, NY · Twitter for iPhone

Prof @covidtweets

And just like that, today's kids will one day grow up and realize the atrocities we committed against them, because we were terrified of a virus that we knew from day 1 was harmless to them.

I hope they make us pay...

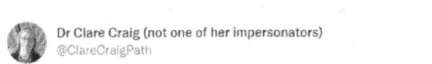
> **Vanessa** @vlal42 · Oct 25, 2021
> At dinner tonight my 15 yo daughter said, "I need to really process what happ in 2020. It really messed up parts of my childhood." 😢 😞
> Show this thread

8:54 PM · Oct 25, 2021 · Twitter for iPhone

Dr Clare Craig (not one of her impersonators) @ClareCraigPath

There's a lot we can learn about SARS-CoV-2 by studying influenza.

The key part of the puzzle being why it was around for so many months before surging in March 2020.

7:02 AM · Oct 7, 2021 · Twitter Web App

AJ Kay @AJKayWriter

Vaccines don't stop transmission ergo vaccine passports won't stop transmission.

This isn't astrophysics.

5:31 PM · Oct 2, 2021 · Twitter for iPhone

Anish Koka @anish_koka · Oct 17, 2021

If you can't trust the Cdc/public health/MSM to communicate about issues related to vaccines in young boys, why trust their analysis on anything?

Here's a visualization of just how deep the hole public health has dug itself..

Bachman @ElonBachman

As a meme, "Let's Go Brandon" is beautifully evolved to make the regime sweat

It's wholesome, which makes it uncensorable

And it holds regime propaganda up for ridicule

It's a wink from all to all

5:39 PM · Oct 14, 2021 · Twitter Web App

Tracy Høeg, MD, PhD @TracyBethHoeg

I'm sad this is real
How many childhoods does one get?
In our Wood Co WI study kids ate indoors & *talked* & went to recess *w/o masks*
Now we have vaccines for adults-why are rules for kids getting stricter?
I'm curious who feels this is a good policy 🤔

> toronto.ctvnews.ca
> Toronto students told not to speak during lunch to reduce t...
> Parents of elementary school children in Toronto are expressing concern after their children were asked not to ...

7:53 PM · Oct 24, 2021 · Twitter Web App

Show Me The Data @txsaith2o

If your elementary school child is overjoyed at the thought of getting vaccinated, it's because you've been a terrible parent.

9:21 AM · Oct 30, 2021 · Twitter Web App

Aaron Kheriaty, MD @akheriaty

Any society that uses children to shield adults from harm has entirely lost its moral bearings.

11:30 PM · Oct 26, 2021 · Twitter Web App

TEAM REALITY

Phil Holloway™ 🌿 🐘 ⚔
@PhilHollowayEsq

"Covid Exposure Anxiety" will be the next disorder added to the DSM. It's vastly more prevalent than #COVID19. It spreads very easily, is manifested by irrational paranoia, an authoritarian mindset, compulsive PCR testing, and debilitating germaphobia.

11:19 AM · Oct 16, 2021 · Twitter Web App

Eli Klein
@TheEliKlein

Direct quote from today's The NY Times

"In The New York Times, an epidemiologist predicted that cases would rise in September because children were going back to school. And what actually happened? Cases plunged."

Remember this when "experts" blame schools for Covid in winter

7:41 AM · Oct 8, 2021 from Manhattan, NY · Twitter for iPhone

Matt, Pre-School Diploma 😀
@statomattic

Obesity is healthy! Masking kids actually helps them learn! One-size-fits-all cloth stops a virus!

Seriously? When people look back at this era in 50 years, they are going to see us as the dummiest dummies who ever dummied.

11:11 AM · Oct 14, 2021 · Twitter for Android

Show Me The Data
@txsalth2o

If you think your kids don't mind wearing a mask, tell them they don't have to wear one anymore.

Watch the reaction.

4:52 PM · Oct 14, 2021 · Twitter Web App

Angry Cardiologist
@AngryCardio

When the CDC adjusts its COVID vaccine recommendations in light of myocarditis risk among young men/boys following 2nd mRNA dose, what will be the reaction of docs who publicly approved current, out-of-step recommendations?

"Knew it all along."	14.1%
"Science evolves…"	70.6%
"Big mistake, CDC!"	15.3%

3,419 votes · Final results

10:40 PM · Oct 6, 2021 · Twitter for iPhone

Vanessa
@vlal42

It is NOT "just" a mask. It is an impediment to hearing, seeing smiles/emotional facial cues, rapport building. It's uncomfortable to wear one for hours on end & exacerbates acne & dry, chapped skin. For kids, it is 1 more layer that impedes connections. It is NOT "just" a mask.

12:57 PM · Oct 29, 2021 · Twitter Web App

Dr Clare Craig (not one of her impersonators)
@ClareCraigPath

Why did nearly a million people who did not have COVID get tested yesterday? It is madness.

Everyone I know who has had COVID could have diagnosed themselves without a test.

The lies about asymptomatic disease and it presenting like a common cold have caused untold damage.

6:27 AM · Oct 19, 2021 · Twitter Web App

AJ Kay
@AJKayWriter

My daughters' school's Covid policy?

'If you're sick, stay home.'

That's pretty much it,

In 9 weeks, guess how much educational disruption, serious illness, absenteeism, and hysteria we've seen?

None. Zero. Nada.

(No masks either.)

5:22 PM · Oct 19, 2021 · Twitter for iPhone

Stinson Norwood
@snorman1776

From 2010

wsj.com/articles/SB100…

11:53 AM · Oct 22, 2021 · Twitter for iPhone

TEAM REALITY

Matt, Pre-School Diploma 🎓
@statomattic

Stop saying "virtual." Start saying "fake."

Stop saying "asymptomatic." Start saying "not sick."

Stop saying "remote learning." Start saying "unofficial school closure."

Stop saying "new normal." Start saying "moral panic."

Change the language, and you change the dialogue.

5:04 PM · Oct 5, 2021 · Twitter for Android

ClownBasket
@ClownBasket

My daughter and I went to an NFL game yesterday. Enclosed space. 64,000 unmasked.

Today my daughter is in school. Masked.

9:24 AM · Oct 4, 2021 · Twitter for iPhone

BOUTROS 木
@boutros555

Why does @CDCgov expect children who can't tie their own shoes to wear PPE correctly for six to eight hours per day?

10:45 AM · Oct 21, 2021 · Twitter Web App

Ian Miller
@ianmSC

For nearly 15 months, Sweden has reported fewer population adjusted deaths than Germany, despite Germany's overwhelming "higher grade" mask usage, and their endless "mitigations" and restrictions

I wonder why Sweden doesn't get praised more by the media for their COVID response

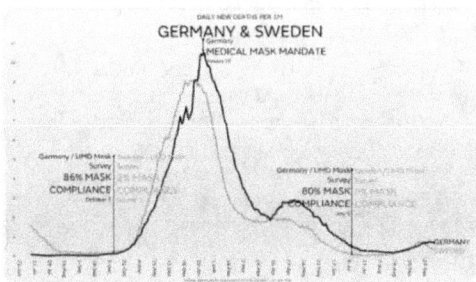

10:49 AM · Oct 13, 2021 · Twitter for Mac

Prof Francois Balloux
@BallouxFrancois

The moral panic during the HIV/AIDS epidemic in the 1980s fuelled enormous homophobia. I find it troubling that largely similar dynamics seem at play during COVID-19, just scapegoating different sections of society.

4:17 AM · Oct 19, 2021 · Twitter Web App

AJ Kay
@AJKayWriter

Humans coexist with a vast bacterial/fungal/viral microbiome with which we have coevolved for millennia.

Sharing "germs" is not a moral failing - it's a precondition for life.

11:21 AM · Oct 8, 2021 · Twitter for iPhone

Anish Koka
@anish_koka

If California and Florida now have a similar age adjusted COVID mortality, is it not eminently reasonable to ask what exactly California's lockdowns were for?

9:47 PM · Oct 10, 2021 · Twitter for iPhone

Martin Kulldorff
@MartinKulldorff

Key priorities for public health agencies:
1. Return to basic principles of public health
2. Apologize for the biggest public health fiasco in history
3. Rebuild public trust

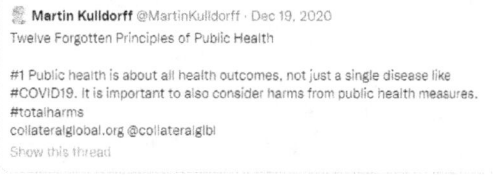

10:41 AM · Oct 7, 2021 · Twitter Web App

Ian Miller
@ianmSC

Not only are cases in Vermont still rising, just the other day they reported their highest single day total of the pandemic, with 90% of 18+ at least partially vaccinated

You really have to hand it to Fauci, he gets essentially everything wrong & never has to answer for it

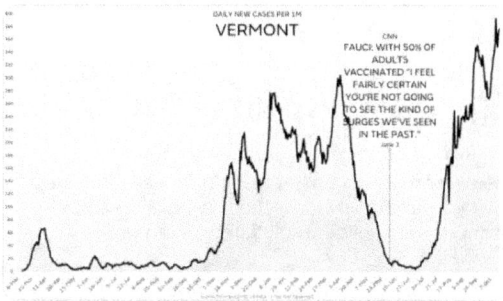

5:16 PM · Oct 20, 2021 · Twitter for Mac

TEAM REALITY

Prof Francois Balloux
@BallouxFrancois

I suspect those overhyping the efficacy of covid19 vaccines, in particularly against transmission, and dismissing any rare side-effect, may have done as much to fuel vaccine hesitancy than ideological vaccine opponents spouting bizarre nonsense (5G, syncitin-1 or whatnot).
1/

10:36 PM · Oct 24, 2021 · Twitter Web App

Worse, I suspect the two groups actually synergistically contribute to sowing doubts about vaccines in the population. ++ For example, if senior academics claim vaccines block viral transmission, and people learn about others having been infected post-vaccination, they will rightly conclude this was untrue, which will provide ammunition to those claiming vaccines don't reduce transmission at all. ++ It would seem ideal if everyone tried to stick to the data. Current covid19 vaccines are excellent at reducing morbidity/mortality upon infection, they reduce transmission to an extent, and have limited side-effects. Their benefit is maximal to those most at risk to covid19. ++ I don't believe it is difficult to make a factual case for covid19 vaccination. There is extensive unassailable and striking evidence that they reduce deaths. Example below: Bulgaria and Romania have the lowest vaccination rates in the EU (~20% and ~30% respectively). ++ Fair enough, "correlation is not causation" and we cannot formally rule out that the high rates of covid19 deaths in Bulgaria / Romania are not caused by low vaccine coverage, but, say, by high cabbage consumption. Thus, some data from the UK below.

Prof
@covidtweets

Find the date when the schools started on this graph. With a ban on mask mandates...

Are they ever going to admit it?

Daily New Cases in Florida

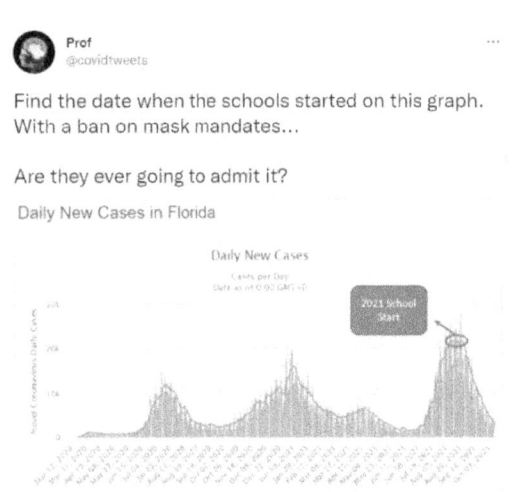

12:22 AM · Oct 12, 2021 · Twitter Web App

Josh Stevenson
@ifihadastick

When the New York Times allows their primary Covid beat reported to overstate the hospitalizations of kids by 14X before quietly correcting, you don't need to wonder why peoples perception of risk to kids is skewed.

🔵 Phil Kerpen @kerpen · Oct 7, 2021
I see this NYT reporter is meeting her usual standards today.

Bethany S. Mandel
@bethanyshondark

Vaccinate your children with a vaccine the White House has already decided the FDA will approve. But there will be no difference in how we handle this virus even if your kids are vaccinated. They'll still have masks, etc. Really making a great sell.

10:42 AM · Oct 20, 2021 · Twitter for iPhone

Ann Bauer
@annbauerwriter

I'm stuck in a city where I know no one, rental cars are sold out, hotels are full, and I may not be able to fly home due to the airline workers' strike against govt mandates.

I'm ELATED. I will take an uber, find a Red Roof Inn and crawl home if I have to. Keep the faith. 🙌❤️

10:48 PM · Oct 10, 2021 · Twitter Web App

Sarah Beth Burwick
@sarahbeth345

A longtime friend told me last week that my tweets are "embarrassing and disturbing."

I am embarrassed and disturbed that so many parents are ignoring the harms we are inflicting on this nation's children.

I will never apologize for advocating for my kids.

10:30 AM · Oct 27, 2021 · Twitter for iPhone

TEAM REALITY

Nick Foy @TheNickFoy

Some other, more impactful ways to protect kids:

-Put them in swim lessons
-Let them play for at least an hour a day
-Encourage healthy eating
-Drive carefully
-Show them that they're loved
-Focus on protecting them from things that are actually bad for them

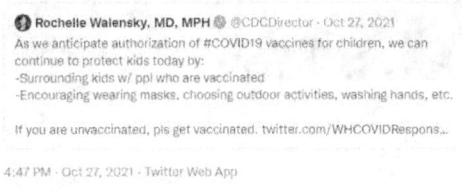

4:47 PM · Oct 27, 2021 · Twitter Web App

Brumby @the_brumby

Labor shortages, supply disruptions, energy inadequacy, & inflation are hallmarks of socialist economies

In Mar '20 a group of idiots halted centuries of market efficiency in favor of a planned economy

We'll be living with the consequences of their stupidity for quite some time

5:50 PM · Oct 13, 2021 · Twitter Web App

Michael P Senger @MichaelPSenger

DeSantis: "We've gone from 15 days to slow the spread to 3 jabs or lose your job."

5:09 PM · Oct 28, 2021 · Twitter Web App

John Ziegler @Zigmanfreud

In September, we saw schools reopen in-person & football stadiums packed across the country...

And yet, somehow, we are averaging significantly fewer new cases on 10/1 than we were on 8/31.

How the hell is that possible, unless the experts were totally wrong about SO very much?!

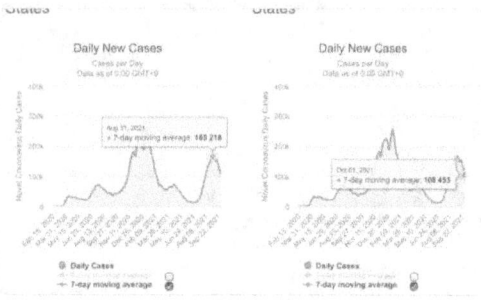

11:38 PM · Oct 1, 2021 · Twitter for iPhone

Emma Woodhouse @EWoodhouse7

We are now at three months of in-person classes at mask-optional schools across the country.

What's the body count on the number of children killed by these millions of naked faces?

3:38 PM · Oct 29, 2021 · Twitter for iPhone

Pajamas It Is @HeckofaLiberal

Well, now you know.

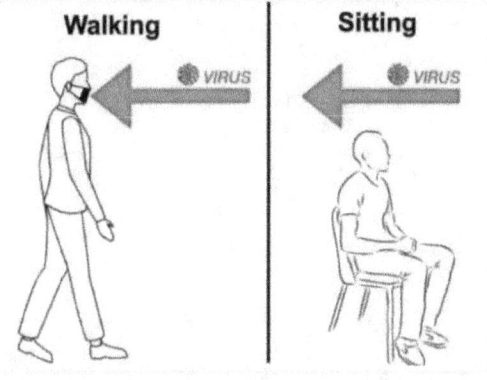

8:05 PM · Oct 14, 2021 · Twitter for iPhone

Aaron Kheriaty, MD @akheriaty · Oct 26, 2021

Let the record show: I am convinced that today's decision by the FDA to give EUA authorization for 5-11 y.o. children for the Pfizer vax will be seen as one of the worst medical disasters of the 21st Century. Say what you want in response, and we'll check back in 20 years.

TEAM REALITY

Sarah Beth Burwick
@sarahbeth345

I have spent the last few months engaging in spot fact checking items I see posted by left wing accounts I have followed for years. And all I can say is, wow.

It's refreshing and liberating to look at multiple views. But it's scary that it took me this long.

1/2

2:35 PM · Oct 28, 2021 · Twitter for iPhone

Emma Woodhouse
@EWoodhouse7

Back to doing this

10:11 PM · Oct 21, 2021 · Twitter for iPhone

Jennifer Cabrera
@jnaskinscabrera

Even if you have gotten the shot, you should be standing with people at your company who are faced with losing their jobs. Tell your employer that you will quit if anyone is fired. This only stops when everyone stands together.

11:20 AM · Oct 28, 2021 · Twitter Web App

Bethany S. Mandel
@bethanyshondark

Can the airline industry please start advocating for a drop of the mask mandate? They have such a nightmare on their hands with noncompliance and they're losing money from families like mine. We have young children and do not want to run the risk of getting kicked off a flight.

5:20 PM · Oct 15, 2021 · Twitter for iPhone

Gummi Bear
@gummibear737

Vaccine Mandates for Seasonal Respiratory Virusus is Authoritarian

Vaccine Mandates for Seasonal Respiratory Viruses in Children is Evil

6:59 PM · Oct 12, 2021 · Twitter for iPad

Eric
@The_OtherET

I'm not a "public health expert" but I'm pretty confident the goal of public health isn't to stretch the truth, misrepresent data and demonize groups of people to scare and coerce them into doing what you want.

1:12 PM · Oct 25, 2021 · Twitter for iPhone

Emily Burns 😊 DMs welcome #TeamReality
@Emily_Burns_V

1/
Just looked @ CDC's AZ school-mask study. What a load of garbage. Here's why:

-No info on actual # of cases, or # of kids in school
-No info on testing levels
-52% of schools WITH mask reqs were small (<850) vs. 13% in No-Mask Schools
-Case rates 2.4x in no-Mask Areas

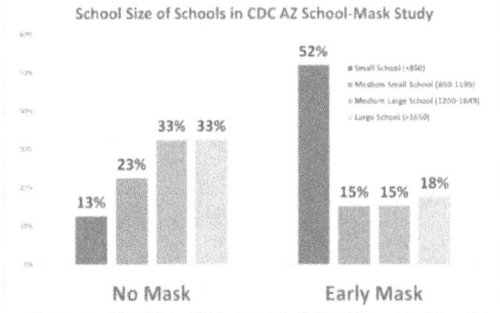

12:59 PM · Oct 7, 2021 · Twitter Web App

Emily Burns 😊 DMs welcome #TeamReality
@Emily_Burns_V

2/
When numbers are missing, it tells you something. The key number here SHOULD be, number of cases/child. That they chose outbreaks instead is...fishy. That 52% of masked schools were small, vs. 13% of un-masked, is important. Fewer kids in schools = Outbreak less likely.

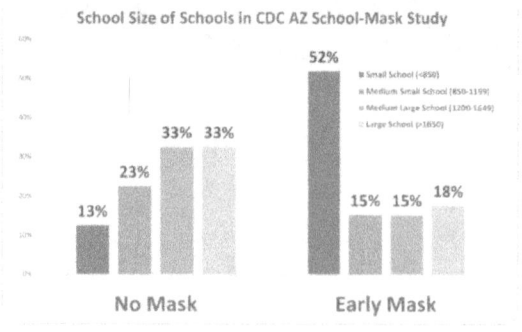

1:02 PM · Oct 7, 2021 · Twitter Web App

TEAM REALITY

 **Emily Burns** 😷 DMs welcome #TeamReality
@Emily_Burns_V

Supposedly the un-vaxed are holding back our recovery.

In truth: the 10 most-vaxed states have ~50% ⬆️er unemployment than the 10 least-vaxed.

10 most-vaxed have 4x ⬆️er incremental unemployment, vs. 10 least-vaxed.

It is the safetyism driving vax mandates that holds us back.

[table image]

2:47 PM · Oct 9, 2021 · Twitter Web App

 **Eric**
@The_OtherET

I just want one person who says we "eliminated the flu" with masks/distancing to explain how Sweden and Brazil managed to eliminate it too, or why most flu seasons originate in east Asia where mask wearing was prevalent pre-2020, or why all the RCTs show masks don't work for flu

9:03 AM · Oct 20, 2021 · Twitter for iPhone

 **elizabeth bennett**
@ebennett74

For all the parents of 5-year-olds who think that vaccinating their kid is going to get him/her out of Dystopia: the college-kid parents are on Line 1.

10:22 PM · Oct 8, 2021 · Twitter for iPhone

 **Jenin Younes**
@Leftylockdowns1

For the record, in like five years when everyone claims they knew lockdowns were catastrophic, masks stupid, and vaccine mandates counterproductive, I plan to be here for the sole purpose of calling them out

10:49 PM · Oct 28, 2021 · Twitter for iPhone

 **Erich Hartmann**
@erichhartmann

Remember: the LockDowns created these supply chain issues.

Our elite "experts" panicked, threw out our Pandemic Playbooks, and then stupidly threw trillions (we don't have) at the problem they created, resulting in an endless series of unnecessary catastrophes.

10:45 AM · Oct 6, 2021 · Twitter Web App

 **Zac Bissonnette**
@ZacBissonnette

I think the basic problem is that there are two basic truths of covid:

1.) It's a horrific virus that killed more than 700,000 Americans
2.) Covid poses less risk to kids than the flu

And for whatever reason, most people just cannot hold these two things in their minds at once

3:29 PM · Oct 3, 2021 · Twitter for iPhone

 **Zac Bissonnette**
@ZacBissonnette

The result is we've done a terrible job doing things that might have helped—free restaurant food delivered to seniors at home is an obvious policy that took off nowhere in the US.

But we ruined education and social services for poor kids, and for no clear covid benefit.

3:31 PM · Oct 3, 2021 · Twitter for iPhone

NOVEMBER 2021

Stefan Baral ✓
@sdbaral

BY THE TIME YOU CLOSE YOUR BORDERS, THE VARIANT IS ALREADY THERE.

(my first all caps tweet)

10:19 AM · Nov 26, 2021 · Twitter Web App

TEAM REALITY

David M @ComradeDoom1

isn't it wild that there's a city in America in which children are not allowed to show their faces when they play together outside but adults regularly shoot up heroin in broad daylight?

8:31 PM · Nov 10, 2021 · Twitter Web App

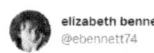
elizabeth bennett @ebennett74

You daily reminder that many respiratory viruses mutate all the time—and in The Before Times, this was never an excuse to take away someone's job, or school, or cover the faces of small children, sigh

8:47 AM · Nov 26, 2021 · Twitter for iPhone

Jennifer Cabrera @jhaskinscabrera

There is no argument whatsoever for discrimination against unvaxed in any setting, even hospitals.

Not only can vaxed transmit COVID, they may be more likely to do so because their symptoms may be suppressed, so they go to work sick without knowing it.

1:21 PM · Nov 13, 2021 · Twitter for iPhone

Don Wolt @tlowdon

We found ants in our pantry again today, which makes no sense because we'd banned ants & had been following a strict ZeroAnt protocol for months. The neighbors tend to be pretty reckless, though, and say things like, "We have to learn to live with ants." This is on them.

12:05 AM · Nov 29, 2021 · Twitter for Android

Dr Clare Craig (not one of her impersonators) @ClareCraigPath

Apparently it would have been age discrimination for the elderly to shield while the young went about their lives.

However, when the unvaccinated are stopped from participating in life, while the elderly are free, that's just fine.

11:35 AM · Nov 16, 2021 · Twitter Web App

Nick Foy @TheNickFoy

If you're crying tears of joy because your kids got their covid vaccine so now they're safe but you don't cry tears of joy every time they arrive at their destination after a car ride you might not be assessing risk very well.

1:40 PM · Nov 14, 2021 · Twitter for Android

Anish Koka @anish_koka

Appreciate authors of the ▇ RCT finally releasing raw data.

Dismayed at their topline conclusion on mask effectiveness that generated so much buzz

Out of ~340,000 ppl in mask and control arm.. the difference in symptomatic cases was 20 over 8 weeks.

> argmin.net
> Revisiting the Bangladesh Mask RCT.
> Musings on systems, information, learning, and optimization.

12:37 PM · Nov 24, 2021 · Twitter for iPhone

Dr Clare Craig (not one of her impersonators) @ClareCraigPath

The Omri-con, Moronic variant, is a psy-op.

No variant has been much different to what went before.

They have each been more transmissible at the beginning and more deadly at the peak. Comparing its progress to the average for the previous entire wave is not a fair comparison.

3:45 PM · Nov 28, 2021 · Twitter Web App

BOUTROS 木 @boutros555

Being on the right side of history often means being on the wrong side of the present day.

11:50 AM · Nov 14, 2021 · Twitter for iPhone

Brumby @the_brumby

1 in 13 deaths from Swine Flu were in children. We did nothing.

1 in 1300 deaths *with* COVID are in children. We ruined their lives.

2:58 PM · Nov 5, 2021 · Twitter Web App

Real Developments @pdubdev

Taiwan halts Pfizer for kids because of high myocarditis risk.

Europe halts Moderna for young people because of myocarditis

America plunges on

11:37 PM · Nov 10, 2021 · Twitter Web App

TEAM REALITY

Tracy Hoeg, MD, PhD
@TracyBethHoeg

Since the school year's start, @ordinrysprheros , @ifihadastick & I have been following 2 k-12 public school systems in Fargo, Cass County, ND, both w/around 12,000 students
Prelim data:
FPS (blue)= mask mandate
WF (yellow)= masks optional
We hope @CDCgov will be interested, too

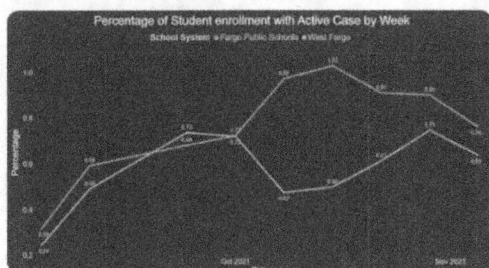

10:08 AM · Nov 7, 2021 · Twitter Web App

Mark Changizi
@MarkChangizi

You're on the wrong side of an issue if your side abandons any of the following:
~ civil liberties
~ cost-benefit analyses
~ free expression
~ informed consent

The Covid Cult has abandoned all of them.

1:02 PM · Nov 20, 2021 · Twitter Web App

Phil Holloway™
@PhilHollowayEsq

#COVID19 restrictions and mandates all begin with the false premise that a common, widespread respiratory virus can be controlled in any meaningful way. Proponents of such also fail to consider the harms to society and overall health associated with these abnormal measures.

5:01 PM · Nov 21, 2021 · Twitter for iPhone

Megan "Hey it's the mask lady" Mansell
@mamasaurusMeg

So I have to say it.

Omicron..

.06 microns

Are we just going to refuse to see viruses as a particulate forever? Because that's where risk mitigation begins and fearmongering ends.

People act like a virus is a shapeshifter chasing them through walls and across vast distances.

11:41 AM · Nov 27, 2021 · Twitter for iPhone

Tracy Hoeg, MD, PhD
@TracyBethHoeg

20% of kids being excluded from public education in LA would be devastating & regressive is exactly the right word. Remember, this vax does not completely prevent transmission, is still new & should be by choice. Public health should be about supporting, not punishing children.

> Vinay Prasad, MD MPH @VPrasadMDMPH · Nov 22, 2021
> I again urge LA & other school districts to not carry out this regressive policy
>
> I explained why in USN&WR: The penalty is too harsh for kids who already lost more than a year
> usnews.com/opinion/articl...

11:37 PM · Nov 22, 2021 · Twitter Web App

Phil Kerpen
@kerpen

OUT: Masks as source control.
IN: Masks as punishment and social ostracism for choosing not to get a vaccine.

Biden's OSHA mandate is even worse than previously reported.

osha.gov/sites/default/...

testing for employees who are not fully
d. The ETS requires employers to ensure
employee who is not fully vaccinated is
COVID-19 at least weekly (if in the
at least once a week) or within 7 days
turning to work (if away from the
for a week or longer). The ETS does not
mployers to pay for any costs associated
ng. However employer payment for
ay be required by other laws, regulations,
ive bargaining agreements or other
ly negotiated agreements. In addition,
rohibits employers from voluntarily
the costs associated with testing.

Face coverings. The ETS requires employe
ensure that each employee who is not full
vaccinated wears a face covering when in
when occupying a vehicle with another pe
work purposes, except in certain limited
circumstances. Employers must not preve
employee, regardless of vaccination status
voluntarily wearing a face covering unless
a serious workplace hazard (e.g., interferi
the safe operation of equipment).

8:53 AM · Nov 4, 2021 · Twitter Web App

Pajamas It Is
@HeckofaLiberal

Tonight MSNBC claimed Republican victories were because of racism, as a Dem I can promise you it wasn't that at all. It was bad covid policies and mandates. People want their lives back and Dems hold us hostage.

5:40 AM · Nov 3, 2021 · Twitter for iPhone

Eli Klein
@TheEliKlein

The stock market's fall today had little to do with how dangerous this new Covid variant is, and everything to do with how dangerous the world's misreaction to it may be.

3:56 PM · Nov 26, 2021 from Manhattan, NY · Twitter for iPhone

TEAM REALITY

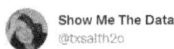
Show Me The Data
@txsalth2o

We are no longer democrats or republicans.

You're either for freedom, or tyranny.

Everything else is insignificant.

12:16 PM · Nov 9, 2021 · Twitter Web App

Eli Klein
@TheEliKlein

We're weeks away from blue state Covid restrictions that won't be implemented because they work, but because they must be seen as doing something. And no matter how bad it gets, they can always say it would have been worse if they didn't perform pseudoscience. A vicious cycle.

11:12 PM · Nov 22, 2021 from Manhattan, NY · Twitter for iPhone

Wes Pegden
@WesPegden

It seems unlikely that future students of public health will admire our multiyear persistence maintaining a singular, laser-like, often sensationalist focus on one very particular threat to human health to the detriment of nearly all other human endeavors and aspects of wellness.

6:59 PM · Nov 28, 2021 · Twitter Web App

Eric
@The_OtherET

A nutritionist that Twitter promotes as a COVID expert suggests masks reduce transmission by 98% and the director of the CDC says masks protect the wearer by 80%. Both are objectively false in light of real-world data

Since they have credentials I guess this isn't misinformation

4:27 PM · Nov 5, 2021 · Twitter for iPhone

Stinson Norwood
@snorman1776

It's odd to see people refusing to give up the ghost since all of this has happened before. Even Tony knew it.

'Fauci said pandemic flu such as H1N1 tend to push out seasonal flu strains. "If we get deluged with H1N1 flu, there is a good chance it will crowd out seasonal flu."'

11:24 PM · Nov 15, 2021 · Twitter Web App

Angry Cardiologist
@AngryCardio

Many folks seem surprised when I say that we are unlikely to understand the mechanism of mRNA vaccine-associated myocarditis anytime soon.

For those who think we can, please outline what experiments you will do, and how they will help elucidate the mechanism.

9:45 PM · Nov 6, 2021 · Twitter for iPhone

Jay Bhattacharya
@DrJBhattacharya

The school closures will catapult wider race and income inequality into the next generation.

Of the many egregious blunders the lockdowners and Dr. Fauci made during the pandemic, pushing to keep schools closed could be the single worst.

11:10 PM · Nov 5, 2021 · Twitter Web App

Stinson Norwood
@snorman1776

9:27 AM · Nov 8, 2021 · Twitter for iPhone

BOUTROS 木
@boutros555

Sometimes I just sit back & contemplate how large swaths of America are living completely normally & paying no epidemiological price, while the deep blue parts keep panicking & self-flagellating to no epidemiological gain.

It's like some places just want to be miserable.

8:38 PM · Nov 26, 2021 · Twitter Web App

TEAM REALITY

AJ Kay
@AJKayWriter

I'm on an @AmericanAir flight as I type & the crew has spent 90% of the time on the intercom delivering repeated warnings about letting masks slip, including a stern rebuke that two reminders will result in authorities meeting you at the gate & inclusion on a no-fly list...

2:40 PM · Nov 14, 2021 · Twitter for iPhone

What I don't get is that all but the small, fearful minority know that masks are a mere performance whose only utility is *placating the feelings* of an emotionally dysregulated subset of the population... ++ which includes people with strong, self-righteous political/tribal allegiances, people who overestimate the actual risk from Covid, and people who are inclined - for whatever reason - to be society's self-appointed hall monitors... ++ But if the point is to assuage feelings, what about the feelings of those forced to cover their faces against their will? Those required to pretend they are sick when they're not? Those who feel degraded and unseen by such an inhumane and disconnected policy?... ++ People who understand - viscerally and acutely - exactly what is lost by requiring everyone to participate this kind of dehumanizing, pointless, pseudo-religious ritual in order to exercise the human right to move freely?... ++ Those are the people who have a legitimate case to make about feelings. And if you're going to prioritize one group's feelings over another, are you sure you're picking the right one?... ++ You're not going to find the answer in RCTs or data. You'll find it accumulating in the form of disgruntled employees, dissatisfied customers & general human misery. I hope someone in your org reads this and understands the significance of what you are actively co-signing... ++ You are buying into COVID- avoidance as the organizing & primary principle of your product & it is not a path to growth or progress, for business or greater humanity. @MartinKulldorff said that the Enlightenment has ended with our reaction to Covid and I can't help but agree. ++ And @americanair - you, and all the other commercial airlines, are not only complicit. You're culpable. ++ ETA: They also encouraged passengers to narc on their neighbors. They literally said, "If you see a passenger with safety violation, please report it to your nearest flight attendant." ++ ETA: I get the rationale but I won't stop flying. I *will* keep challenging mask & vaccine mandates. My travel isn't frivolous, I travel to meet in-person (critical distinction), learn from, & collaborate w/ ppl who, like me, are trying to right the ship. I won't stop.

David M
@ComradeDoom1

gotta say, every time a covid Expert™ compares masks to seatbelts, I become ever so slightly skeptical of seatbelts.

10:01 PM · Nov 18, 2021 · Twitter Web App

Prof Francois Balloux ✓
@BallouxFrancois

"Covid's law"

Whenever a consensus emerges that a country got its pandemic policies right, things get hairy in said place.

3:46 AM · Nov 8, 2021 · Twitter Web App

BOUTROS 木
@boutros555

So let me get this right: The "civilized" world enacted travel bans due to a variant that's already prevalent everywhere that we only thought to investigate because a physician noticed that her COVID patients were exhibiting *milder* symptoms than usual.

Is that the gist of it?

4:46 PM · Nov 29, 2021 · Twitter Web App

Stinson Norwood
@snormant776

You know the Great Barrington Declaration and the John Snow Memorandum, but do you recall the most (not) famous open letter of all?

Endorsed by the New York Times in early March, 2020; signed by almost 800 experts and entities. The Gonsalves Letter.

Gregg Gonsalves @gregggonsalves · Mar 6, 2020

6:10 PM · Nov 26, 2021 · Twitter Web App

TGL: "Government...must recognize that low-wage, gig-economy, and non-salaried workers who are unable to work because ... disruptions to the economy and public life face extraordinary challenges. They may find it impossible to meet their basic needs, or those of their family." ++ GBD: "Keeping these measures in place until a vaccine is available will cause irreparable damage, with the underprivileged disproportionately harmed." ++ GBD: "Simple hygiene measures, such as hand washing and staying home when sick should be practiced by everyone to reduce the herd immunity threshold." ++ TGL: "Infringements on liberties need to be proportional to the risk presented by those affected, scientifically sound, transparent to the public, least restrictive means to protect public health, and regularly revisited to ensure that they are still needed as .. epidemic evolves" ++ TGL admits the virus is gonna virus: "Efficiently identifying those exposed will be increasingly difficult as community transmission of the virus becomes more widespread, making quarantine a less plausible measure as community spread proceeds." ++ So does the GBD: As immunity builds in the population, the risk of infection to all – including the vulnerable – falls. We know that all populations will eventually reach herd immunity" Note the word "eventually." There isn't a suggestion to lick doorknobs in the GBD. ++ TGL is very much on board with maintaining liberty. There is no mocking of "freedumb" in this letter. ++ TGL: "Healthcare workers must...be protected from discrimination arising out of their work with infected patients." Very odd that we want to fire so many of them now. Even more so since @doritmi also signed The Gonsalves Letter. ++ A great idea from TGL: The GBD mentioned using recovered employees to reduce the strain. Both great ideas. Does anyone know if we focused much effort on capacity of our healthcare systems? Any of the CARES money go towards expanding capacity? ++ The Gonsalves Letter, just like the Great Barrington Declaration, derive from many traditional public health concepts and pandemic planning. ++ So what changed? TGL was written around 3/4/20. At that time there were rumors of a federal shutdown. Andrew Cuomo: "You don't want to shut down society, because that's massively disruptive — to the economy, to life" ++ During a Cuomo Bro discussions, the elder Cuomo responds to some of these rumors and sounds positively like a "Covid minimizer" ++ The same day as that interview VP Pence was printing off "15 days to slow the spread" placards. So what changed? How did the GBD publish an updated open letter, with fresh epidemiological data and traditional public health concepts, and get pilloried, just a few months later?

TEAM REALITY

>  **Matt, Pre-School Diploma** 🎓
> @statomattic
>
> With the covid debate now mostly centered around mask mandates, vaccine mandates, and schools, we must not forget, lockdowns were the original sin that made all the rest possible. 🧵 (1/)
>
> 8:52 PM · Nov 17, 2021 · Twitter for Android

Without the original, fraudulent, 15 days to flatten the curve, businesses would not have been forced to close, thereby avoiding much of the spike in unemployment and - eventually - today's supply chain woes. But that's just on the surface. There's much more. ++ Had we never locked down, people would have kept working and going to school out of the house. They would have maintained a near-typical level of contact with the real world. This is critical, as it enables information inputs other than artificial media. ++ Without lockdowns, the media's coordinated propaganda could not have grown roots as deep as it did. Dr. Fauci would still have been a public authority, but would never become the anointed, demi-God-like, unimpeachable voice of The Science. ++ In those critical weeks of March and April 2020, much of America barely left home save for grocery runs. Shelter from the outside world, plus the fear-stoking media narrative with its "we're all in this together" propaganda, was tremendously effective. ++ Much like the villagers in Shyamalan's "The Village," the sheltered populace was convinced, via propaganda, that "the towns" were dangerous. Strangers, viewed automatically as dangerous and diseased. ++ In light of this view, anyone resisting the popular narrative represented a unique danger. These people weren't merely showing a different assessment of risk, for if we were all in this together, they were NOT. They were callous deniers, sabotaging our lives selfishly. ++ The stage had been set. If Fauci was The Science, if we were all in this together, if the lockdown skeptics were infidels, then any deaths from the virus could be blamed on THEM. Not a virus, but the others. That included the many esteemed scientists who disagreed ++ In this created environment of partisan conflict and fear, augmented by the political bifurcation of present-day America, every point now becomes a political lightning rod. If the righteous support it, it must be mandated. If not, it must be banned ++ And so, a cloth mask - seen in saner times as a largely ineffective blunt tool to perhaps try temporarily out of desperation - took on a religious quality. The Covidian must mask without ceasing, as the Christian is instructed to pray. We take on faith that it works. ++ When the despised President Trump said, in July 2020, that schools must open, closing schools too became a tenet of the faith. The needs of children, cast aside due to the partisan political narrative. Surely you are not a Science-Denying Trumper. ++ Staying home itself became a snobbish, elitist badge of honor. Venturing out of the home for work, taking trips, became the domain of vulgar rubes. The Righteous eloi would order their Doordash and let the morlocks deliver it. Contact and sight-free, of course. Stay safe. ++ In the end stages of the pandemic, even vaccines became, ludicrously, a political lightning rod. A fair assessment is almost impossible, tarred as they are by political religion. You are either vaxxed to the max or a filthy leper. There can be no middle ground. ++ Seriously, what even is this? In our current environment, the vaccines are either an unimpeachable miracle (with the need to boost and mask, illogically) or nefarious poison. It's all or nothing. Except that's childish and wrong. Sober assessments are rare. ++ So, we are left with a toxic mess. We can't know the best course of action because everyone fights all the time and nobody can agree on the discussion. Would we be at this point had we never taken Chinese advice and locked down? I sure don't think so. So, speculation time. ++ Had we never locked down... -Schools would have stayed open -Business would have dipped briefly, then stabilized quickly as panic dissipated -Mask mandates would never have gained much traction -The people would have still been politically polarized, but not over covid ++ -Vaccines would be assessed fairly, and not be a topic of religious fervor. The vulnerable would get them. Many others would not. There would be no character judgment surrounding that choice. They would be viewed as a worthy effort but imperfect. ++ -Natural immunity would be acknowledged. -Public health officials would be viewed as specialists, not unelected pseudo-religious political dictators. -Stupidity would still abound, but sanity would largely prevail ++ -Covid deaths would be about the same as they are now, but minus the CNN tickers, the panic, the monomania, the rancor. Without the psychological damage, the damage to children, without fraying of personal connections and societal bonds. The unquantifiable would be saved ++ If only. Eventually, we will return to full normal. We will not be masked forever, even in schools and on planes and in government buildings. We will heal. But it didn't have to be this way. Lockdowns were the original sin. Pure evil. We must never do them again.

Dr Clare Craig (not one of her impersonators)
@ClareCraigPath

Since summer there have been twice as many covid deaths but seven times as many excess deaths as last year.

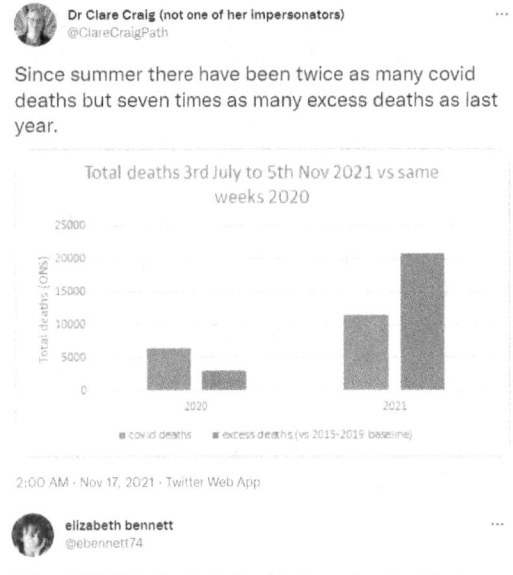

2:00 AM · Nov 17, 2021 · Twitter Web App

elizabeth bennett
@ebennett74

When COVID is finally judged to be endemic rather than pandemic by The Powers That Be, there will be a lot of glossing over what was done to children. "We made the best decisions we could at the time," etc. Do not buy that BS. And don't forget the worst offenders.

11:41 AM · Nov 7, 2021 · Twitter for iPhone

David M
@ComradeDoom1

'We didn't destroy your lives, covid did (albeit only in places governed by Dems, weirdly enough)' - a winning message from the Democrats going forward

10:23 PM · Nov 23, 2021 · Twitter Web App

BOUTROS 木
@boutros555

Why would Ron De Santis do this?

6:47 PM · Nov 28, 2021 · Twitter for iPhone

TEAM REALITY

Martin Kulldorff
@MartinKulldorff

News of new Nu variant, but WHO is jumping the alphabet to call it Omicron, so they can avoid Xi.

1:27 PM · Nov 26, 2021 · Twitter Web App

Hold2
@Hold2LLC

Do not despair, for this is the last gasp of insanity by Gov't and PH leaders worldwide.

Don't be saddened; be emboldened.

Don't retreat; push forward and bring other converts with you.

We end this.

6:15 PM · Nov 27, 2021 · Twitter for iPhone

Bethany S. Mandel ✓
@bethanyshondark

Congratulations to public school unions who got a new variant just in time to extend their Christmas holidays.

8:20 AM · Nov 29, 2021 · Twitter for iPhone

Eric
@The_OtherET

It may take a while, but the assumption of COVID being spread through droplets/fomites and the "interventions" that came along with it (Social distance dots! Plexiglass! Sanitizer! Masks!) will go down as one of the most disastrous public health assumptions in history

1:39 PM · Nov 24, 2021 · Twitter for iPhone

Jennifer Sey ✓
@JenniferSey

My 4 yr old daughter was sick with a bug this week. I did what any parent would have done pre Covid. I held her & slept by her side. What cruelty to isolate a small child when sick. I don't understand how some can recommend this & how any parent can follow through with it.

12:13 PM · Nov 20, 2021 · Twitter for iPhone

Prof François Balloux ✓
@BallouxFrancois

Or less subtly:
Boffin: We found this funny looking viral variant ...
Official 1: Warn the WHO!
Official 2: Nope, stop, remember what happened last time. They cordoned us off, which devastated our economy, and on top we got all the blame despite it being already widespread ...
5/

8:48 AM · Nov 26, 2021 · Twitter Web App

Kyle Lamb
@kylamb8

A fully vaccinated hockey team just had games canceled due to a COVID outbreak where 40 percent of the roster was infected. As usual, healthy athletes not actually severely ill, mind you, I'll, here we still act like vaccines are protecting the workplace from transmission.

5:30 PM · Nov 15, 2021 · Twitter for Android

M_P
@Reroot_Flyover

Has literally any place in the world actually meaningfully increased health care capacity (facilities, staff or both) as part of their response to the pandemic? I can't think of one.

11:43 AM · Nov 8, 2021 · Twitter Web App

Jay Bhattacharya
@DrJBhattacharya

Public health's purpose is to provide people tools for living healthier. The tools necessarily differ based on needs. Correct public health advice for a baby differs from that for an old man.

To pretend that focused protection was impossible was a dereliction of duty.

1:05 AM · Nov 9, 2021 · Twitter Web App

Sarah Beth Burwick
@sarahbeth345

"Open schools" do not include:
- Forcibly covering children's faces
- Missing instruction time every week for a nose swab
- Parents not permitted to enter even the outdoor portion of campus, much less the classroom
- Requiring healthy kids to miss 2 weeks of school at any moment

12:47 PM · Nov 29, 2021 · Twitter Web App

TEAM REALITY

Woke Zombie 🧟
@AWokeZombie

#US #COVID19 Nov. 19 2020 vs. Nov. 19 2021

Pretty amazing the symmetry.

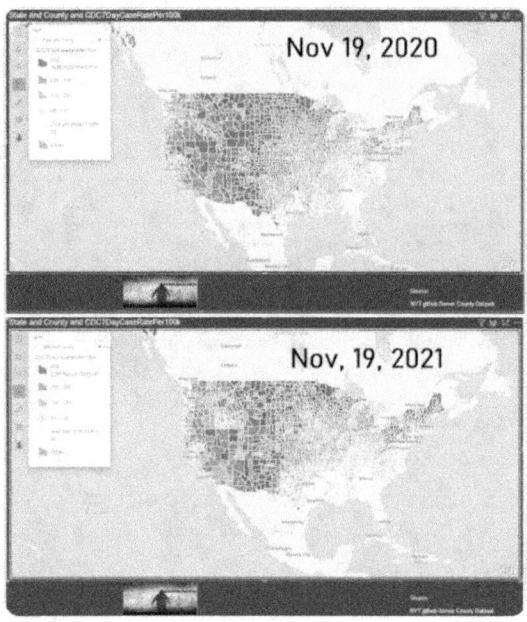

9:26 PM · Nov 19, 2021 · Twitter Web App

Sarah Beth Burwick
@sarahbeth345

Where would we be now if we had vaccinated 60+, focused on treatment options, and ditched all other restrictions?

For real, has anyone modeled this scenario?

5:04 PM · Nov 29, 2021 · Twitter for iPhone

Wes Pegden
@WesPegden

Recently announced travel bans on South Africa and other countries because of the Omicron variant should serve as reminder that our tendency in many cases is not to learn from past mistakes with respect to ineffective policy, but to become desensitized to them and repeat them.

4:57 PM · Nov 26, 2021 · Twitter Web App

Martin Kulldorff
@MartinKulldorff

It's problematic when the same person (Fauci) directs both pandemic policy and the largest funder of infectious disease research (NIAID). Scientists must be able to oppose public health policies without fear of losing research funds.

12:51 PM · Nov 26, 2021 · Twitter Web App

Emma Woodhouse
@EWoodhouse7

Doesn't matter what they call it:

▶ "Switching to Distance Learning"
▶ "Pivoting to Remote"
▶ "Taking an Adaptive Pause"
▶ "Extending the Weekend"
▶ "Adding Days to a Scheduled Break"
▶ "Transitioning to E-Classes"

It's a school closure, & it's hurting kids.

11:23 PM · Nov 23, 2021 · Twitter for iPhone

Gummi Bear
@gummibear737

We should be able to have an honest, open discussion about vaccine injury, without implying that the vaccines don't help people

It's a simple risk/reward analysis

It helps a lot of people based on age, weight and health

It doesn't make sense for my 12/13 year old daughters

1:53 AM · Nov 14, 2021 · Twitter for iPad

Eric
@The_OtherET

Maybe having one of the people most responsible for keeping kids out of school speak at your campaign rally the night before the election was a bad idea idk

11:52 PM · Nov 2, 2021 · Twitter for iPhone

Jay Bhattacharya
@DrJBhattacharya

I am proud to have provided pro-bono expert testimony in DeSantis vs. Florida Education Association last year. The case permitted Florida's schools to stay open all year for in-person instruction over the objection of the Florida teacher's unions.

edca.1dca.org/DCADocs/2020/2...

10:02 PM · Nov 9, 2021 · Twitter Web App

TEAM REALITY

Zac Bissonnette
@ZacBissonnette

It would be cool if Biden or Walensky or Fauci would do a speech and explain what exactly it is we're hoping to accomplish on covid.

3:11 PM · Nov 27, 2021 · Twitter for iPhone

Because these travel bans tear families apart, school systems in blue states aren't functional, and a lot of people are tired and angry. I'm pro-vaccine. Get vaccinated, please! But we need to know what our realistically achievable goal is with all these harmful restrictions. ++ Is our goal to get covid cases down to zero over the next few months? If so, does anyone think that's possible? I know nothing about epidemiology but I do know that policies need to have clear goals—and clear measures of success and failure, and an honest discussion of costs.

M_P
@Reroot_Flyover

BREAKING: Businesses being allowed to do business are doing better than businesses that were not allowed to do business.

> 🏛 **The White House** @WhiteHouse · Nov 16, 2021
> Update:
> Compared to this time last year, retail sales are up 16% and sales at restaurants and bars are up more than 29%.
>
> The Biden economic plan is working.
>
> whitehouse.gov/briefing-room/...

10:18 AM · Nov 17, 2021 · Twitter Web App

Ann Bauer
@annbauerwriter

I will never understand the contorted thinking that goes into: I know this vaccine doesn't prevent infection or transmission of Covid-19, but I will only associate with people who've been vaccinated because its makes me feel safe.

10:18 AM · Nov 23, 2021 · Twitter Web App

Jennifer Sey ✓
@JenniferSey

Losing friends is hard.

Not living in accordance with your values is harder.

5:31 PM · Nov 13, 2021 · Twitter for iPhone

Jenin Younes
@Leftylockdowns1

It is nothing short of astonishing to watch those who for so long proclaimed to be on the side of the American worker now champion employers having absolute control over their employees' personal health decisions.

10:44 PM · Nov 17, 2021 · Twitter Web App

Erich Hartmann
@erichhartmann

NOTE: the same geniuses who are masking our kids at school think this  is "safe" too

9:38 AM · Nov 12, 2021 · Twitter Web App

Prof Francois Balloux ✓
@BallouxFrancois

It might feel paradoxical to be very supportive of vaccines, and to be against vaccine mandates. I don't believe it is, as I worry we may pay dearly in the future any short-term gain in vaccination coverage that was obtained by forcing people to get vaccinated against their will.

9:25 PM · Nov 6, 2021 · Twitter Web App

TEAM REALITY

DECEMBER 2021

Bethany S. Mandel
@bethanyshondark

Nothing has changed except they all got COVID this weekend.

10:08 PM · Dec 19, 2021 · Twitter for iPhone

TEAM REALITY

AJ Kay
@AJKayWriter

Fact: Childhood is finite.

Kids won't get a re-do on almost two years of diluted education, foregone rites of passage, missed opportunities, & a generally muted existence.

The brunt of the burden will fall on kids least equipped to bear it: the poor, the abused, the disabled.

11:59 AM · Dec 9, 2021 · Twitter for iPhone

Bethany S. Mandel ✓
@bethanyshondark

My email to the editor of @Highlights

8:51 AM · Dec 15, 2021 · Twitter for iPhone

Virál Myãlgia MD, PhD @contrarian4data · Dec 29, 2021
I cringe when I see serially wrong media ExpertsTM continue to demonstrate that they do not know the body of literature pre-COVID.

Viral aerosols range in the 0.1um to 0.5um range and take 48h+ to settle. Even properly fitted N95s are rated at 0.3-0.5um.

cdc.gov/niosh/topics/a...

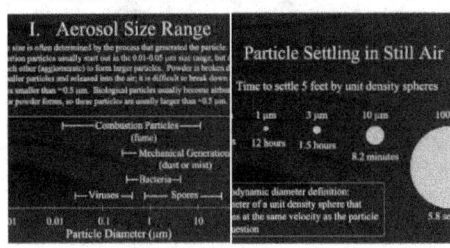

Craig
@TheLawyerCraig

All #Omicron data of which I am aware shows the reduction in severity outpaces the increase in transmission such that severe outcomes will likely be lower *in raw numbers* than other waves (not just as a % of cases).

Please stop talking school closures or lockdowns.

9:50 AM · Dec 20, 2021 · Twitter Web App

Michael Tracey ✓ @mtracey · Dec 27, 2021
Princeton's latest **COVID** policy -- twice-weekly asymptomatic testing, mandatory "boosters," banning gatherings that include food, instructing students to "stay in their rooms as much as possible" -- is set by a person who previously ran the Gender and Sexuality Studies program

ClownBasket 🤡
@ClownBasket

As we get ready for sleep tonight, know that millions of kids will wake up tomorrow prepared to wear a mask for 7+ hours while believing they are a danger to adults.

12:17 AM · Dec 16, 2021 · Twitter for iPhone

David Zweig
@davidzweig

EXPERTS: the CDC study on masks was "so unreliable that it probably should not have been entered into the public discourse"

A multi-month investigation into a flawed study, and a public health agency's lack of transparency

My latest, for @TheAtlantic

theatlantic.com
The CDC's Flawed Case for Wearing Masks in School
The agency's director has said, repeatedly, that schools without mask mandates have triple the risk of COVID outbreaks. That claim is based on very shaky ...

3:57 PM · Dec 16, 2021 · Twitter Web App

Abir Ballan @abirballan · Dec 29, 2021
It looks like SARS-CoV-2 is acting like «other seasonal "common-cold" coronaviruses, which elicit short-term immunity against mild reinfection but longer-term immunity against more severe illness with reinfection»
nejm.org/doi/full/10.10...

Disease Outcome	Reinfection
	no. of persons with o infection that was n
Severe disease	4/1300
Critical disease	0/1300
Fatal disease	0/1300
Severe, critical, or fatal disease	4/1300

Table 1. Severity of SARS-CoV-2 Reinfections as Compared with I

Nick Foy
@TheNickFoy

I'm considering changing careers and entering public health because I think that everyone should do exactly what I want them to do all the time and we should outlaw things that don't interest me.

3:06 PM · Dec 28, 2021 · Twitter for Android

Martin Kulldorff
@MartinKulldorff

As people stop fearing Covid, politicians start fearing people.

6:30 PM · Dec 7, 2021 · Twitter for Android

TEAM REALITY

Clayton P. Cobb
@warrtalon

Virtue or Vice

This is a story about me, my golfing buddy, and our CoV2 debates. I'll call him Mitch.

Mitch is highly-educated, does not trust media, claims to be a political Independent, is inquisitive, and has a lawyer's mind (cross-examine, logical, fact-based).

/1

11:01 AM · Dec 24, 2021 · Twitter for iPhone

For a long time, we never discussed anything related to current affairs (politics, CoV2, etc.). But then I mentioned I was leaving my excellent job of 14 years so that I could stop being anonymous while fighting against CoV2 restrictions, including the local school Board. ++ Being anonymous was holding me back, but I did not want to bring any undue attention to my employer, coworkers, or customers. I also needed to put a lot more time into homeschooling my 3 kids as the primary educator, because we would not allow our kids to be force-masked. ++ Initially, this only spawned conversations of how I would sustain our standard of living: investment types, cash flow, tax-advantages assets, entity structuring, asset protection. But then 1 day, he made an off-hand comment similar to "but I know you don't support vaccines." ++ I quickly replied, "Wait, I do support vaccines but not mandates." This could have been a negative turning point, but thankfully it opened a wonderful door to intelligent, adult dialogue between 2 people who fundamentally disagree. ++ To him, not supporting CoV2 vax mandates was akin to not supporting or believing in the vaccines at all. I explained that vax is a personal choice, should be an informed decision based on individual risk calculus, and should not be coerced. ++ He believed vax mandates were for the greater good and that was enough to override personal choice. He had no empirical evidence of HOW mandates help, but he believed they did simply because "the scientific consensus says they are helpful." This will become a running theme. ++ I showed charts where CoV2 patterns and amplitude moved independent of vax rate and in spite of mandates. I mentioned how HCW mandates only reduce the quantity of staffed beds without improving healthcare (likely a net negative). For example, look at FL vs. NY staffed beds. ++ He argued that most hospitals want the mandates because otherwise they'd lobby hard against mandates. Since they seem to want mandates and most public health experts agree on HCW mandates, then they must be helpful. No evidence, but his logic engine deducts it must be true. ++ We agreed to disagree on HCW mandates, but he did at least say, "you might be right, but I have to go with what the experts say." Not unreasonable. Most people probably feel this way. Except the experts who disagree are vilified, censored, ostracized, and crushed. ++ We also debated masks. He did not know there were already decades of research concluding that community masking of healthy people does not prevent viral (influenza) transmission. Nor that studies since CoV2 were low on the evidence pyramid and were being overstated. ++ He couldn't wrap his head around the possibility that the CDC, WHO, Fauci, and most worldwide Public Health experts might be wrong or might be overstating a useless intervention. This isn't possible, right? Global "consensus" must be correct, right? No. I showed otherwise. ++ He countered with the old PNAS study/model from leading mask proponent Jeremy Howard: https://pnas.org/content/118/4/e2014564118. I explained how it was over a year old and used low-quality evidence from early 2020. A great example is Chechia being included as a top example of mask usage. ++ The others in the study with Chechia were all Oceana countries. I mentioned how it was a lot of early correlation-as-causation observational studies in Oceana that led the West to flip known mask science on its head. We panicked because we had nothing else to fight CoV2. ++ And still to this day, we are clinging to the mask fallacy while forcing toddlers to wear masks on planes and kids of all ages to wear masks for 6+ hours per day in school. All with no evidence of efficacy. I showed the school study results. He had no idea and was shocked. ++ This actually changed his mind, and he no longer agrees with kids being forced to mask in school. His kids are grown, so he had not experienced the many harms. He also had thought school masking reduced community spread because that's what media and Gov't tells us. ++ Though his mind changed for kids, he still chooses to believe Public mask mandates are warranted. His reasoning is there is global consensus that masking helps. Though he now sees the evidence is limited & inconsistent, it's enough for him to "err on the side of caution." ++ Back to vax, boosters, natural immunity, and Omicron. He's vaxxed/boosted and actively attempting to avoid infection still. I showed how that is futile and that it can't be avoided, which means continually imposing restrictions only adds harm and prolongs the process. ++ His belief was that natural immunity was inferior to vax protection and shorter-lived. This belief was a holdover from early studies showing antibodies fall quickly after infection. What he didn't know is they stop falling and remain strong for an indefinite time period. ++ And that vax protection falls off much more quickly regardless of antibody titers, because antibody levels are not predictive for re-infection. These studies show a stark difference with natural immunity being far superior. ++ After those, I showed the recent Vaccine Efficacy (VE) results from Denmark (Omicron vs. Delta) and UK (case rates by age). Notice the NEGATIVE 76% VE vs. Omicron after 90 days. Even 31-90 days shows error bars extending below 0. For UK, vax case rates are higher 18-59. ++ He says hospitals are overwhelmed and HCWs are burned out, so it's justifiable for Gov't to impose restrictions as long as that reduces the # of hospitalizations. Harms to society are outweighed by the need to protect hospitals/HCWs from CoV2 surges. Back to these charts. ++ After all this, his view was definitely shaken a bit, but he concluded with saying he does not have the expertise to make his own informed decisions better than what the experts are telling him. He will continue doing as they say so he can contribute positively to society. ++ We are still good friends, play golf together, eat indoors together (no masks), and respect each other. I believe Mitch represents many Americans who are intelligent, thoughtful, caring, and haven't seen counter-narrative data. Where do you see yourself in this story?

Sarah Beth Burwick
@sarahbeth345

Parents: If you do not think your child is safe at school unless everyone wears PPE, then you really should not be sending the kid to school. Please make alternative arrangements.

8:13 PM · Dec 26, 2021 · Twitter for iPhone

AJ Kay
@AJKayWriter

Fact: Childhood is finite.

Kids won't get a re-do on almost two years of diluted education, foregone rites of passage, missed opportunities, & a generally muted existence.

The brunt of the burden will fall on kids least equipped to bear it: the poor, the abused, the disabled.

11:59 AM · Dec 9, 2021 · Twitter for iPhone

Don Wolt
@flowdon

I created this in September of 2020 and thought, at the time, that the notion of a COVID lockdown continuing all the way to 2022 would be absurd. Heh.

3:46 PM · Dec 28, 2021 · Twitter Web App

TEAM REALITY

Bachman
@ElonBachman

"I got COVID, thank God I was triple vaxxed"

is the new

"I'm not religious, but I am spiritual"

10:47 AM · Dec 20, 2021 · Twitter Web App

102 Retweets   3 Quote Tweets   714 Likes

Michael Tracey
@mtracey

Biden said in July: "You're not going to get COVID if you have these vaccinations." A few media outlets did gently "fact-check" him at the time, but I don't recall a giant freakout about this "misinformation" giving demonstrably false assurances about the efficacy of the vaccine

9:27 PM · Dec 18, 2021 · Twitter Web App

Infectious Disease Ethics
@ID_ethics

If cloth masks are "not appropriate for this pandemic" and "little more than facial decorations"

How could mandating them be ethically justifiable?

Eli Klein @TheEliKlein · Dec 28, 2021
Oh

6:25 PM · Dec 28, 2021 · Twitter Web App

AJ Kay
@AJKayWriter

I wonder if the people who claim we need to "Stop Covid at all costs!" ever stop and think about what those costs might be and who they are nominating to pay them?

1:41 PM · Dec 18, 2021 · Twitter for iPhone

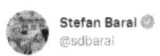
Stefan Baral
@sdbarai

In full disclosure, I have thought vaccine passports were a really problematic idea from the time that I heard the idea floated in April of 2020.

But now, I think they are useless in terms of public health impact too.

1:39 PM · Dec 20, 2021 · Twitter Web App

Ian Miller @ianmSC · Dec 28, 2021
Cases in Washington, D.C. are now twice as high as they've ever been in Florida

I'm still waiting for Vanity Fair to call the D.C. mayor the "Angel of Death" like they did with Ron DeSantis

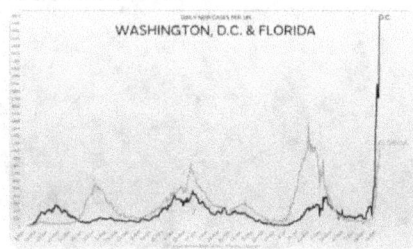

elizabeth bennett
@ebennett74

Healthy people with a sniffle are swamping and overwhelming ERs (including the one where I work). This is squarely the fault of hysterical, fear-mongering media and bad public health messaging.

12:18 AM · Dec 25, 2021 · Twitter for iPhone

Michael P Senger
@MichaelPSenger

CDC Director Walensky on the reduction of quarantine: "It really had a lot to do with what we thought people would be able to tolerate."

This says it all. As much tyranny as the people will tolerate. This ends when we demand they end it—unconditionally.

From The Recount

1:02 PM · Dec 29, 2021 · Twitter Web App

Woke Zombie
@AWokeZombie

#Covid19

NFL coaches and players can come back from Covid in 3 days

HCWs don't even need to stop working if they are covid positive.

But NJ school kids have to quarantine for 7-14 days if they are just NEAR a student who tests positive.

When will enough be enough?

12:51 PM · Dec 26, 2021 · Twitter for iPhone

TEAM REALITY

Kevin McKernan @Kevin_McKernan

Spike Escape and directed evolution.

For those who missed it.

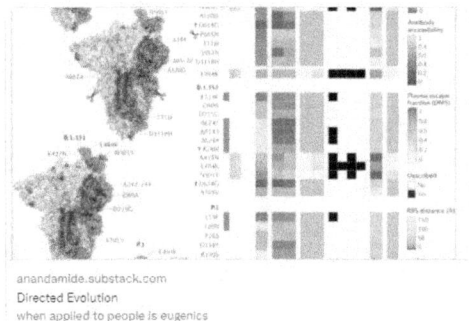

anandamide.substack.com
Directed Evolution
when applied to people is eugenics

6:26 PM · Dec 7, 2021 · Twitter for iPhone

Nick Foy @TheNickFoy

I'm continuing to progress in my public health studies!

Today I learned that we should require everyone who steps on an airplane be vaccinated to stop the spread, but that a college full of vaccinated people shouldn't meet in person because vaccines don't stop the spread.

12:50 PM · Dec 30, 2021 · Twitter Web App

Martin Kulldorff @MartinKulldorff

In ~500 BC, we learned that the earth is not flat, but a sphere.

In 430 BC, during the Athenian plague, we understood natural immunity.

Now universities only believe in one of the two.

8:06 PM · Dec 20, 2021 · Twitter Web App

Tracy Hoeg, MD, PhD @TracyBethHoeg

➕Finland officially not recommending vax for kids 5-11 w/o risk factors
"Infections in children of this age are usually mild & severe disease is extremely rare. When the burden of disease is small in one's own group, very few adverse effects are accepted"
thl.fi/en/web/thlfi-e...

10:02 PM · Dec 5, 2021 · Twitter Web App

Kyle Lamb @kylamb8

If a variant is really transmissible, and if you continue to test every single hospital admission for it, and you continue to count every admission in the statistics regardless of the reason they are there, then it's common sense hospital admissions will rise even if mild.

10:27 AM · Dec 25, 2021 · Twitter for iPhone

Aaron Kheriaty, MD @akheriaty

Fellow physicians: if you are afraid to lose your license or hospital privileges for asking questions or saying what you think, ask yourself: what has happened to our profession? And who do you serve?

9:48 PM · Dec 25, 2021 · Twitter for iPhone

Dr Clare Craig (not one of her impersonators) @ClareCraigPath

Over a year ago someone I love asked me, "Why are you doing this Clare. What do you gain from it?"

I have only lost financially, and in terms of career and reputation in the last two years.

I am doing it so my children have a future.

3:00 AM · Dec 12, 2021 · Twitter Web App

Justin Hart @justin_hart

Their messaging can't get any worse right? Oh no... it can.

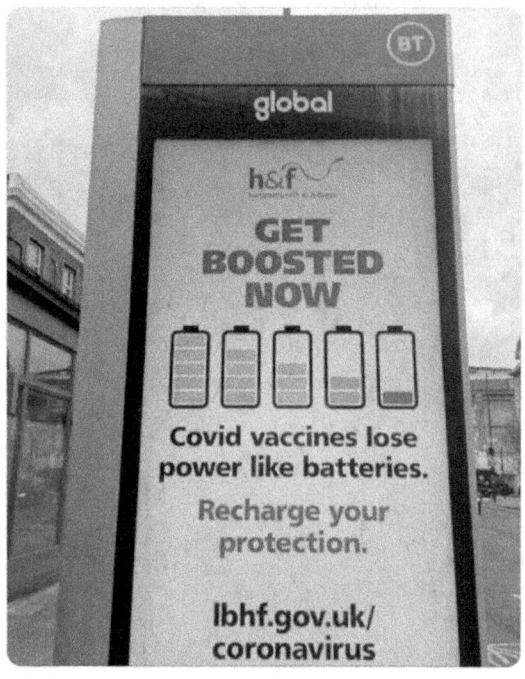

12:20 AM · Dec 24, 2021 · Twitter for iPad

Jay Bhattacharya @DrJBhattacharya

Lockdowners lament that we do not live in an authoritarian society focused on the control of a single infectious disease. Normal people fear that we do.

12:30 PM · Dec 31, 2021 · Twitter Web App

TEAM REALITY

Prof Francois Balloux
@BallouxFrancois

This is not an easy message to convey, even to those who have have already accepted that zero-covid was toast. Essentially everyone will eventually get infected by SARS-CoV-2 in the near future, and likely more than once in their lifetime.
1/

3:41 PM · Dec 27, 2021 · Twitter Web App

One can don an FFP2, FFP3, N95 mask, or a hazmat suit, or whatnot, but at this stage, all this may achieve is to delay the time until some of us will get infected, and thereby marginally prolong the pandemic. ++ Vaccine protection against infection is meh, though protection against severe symptoms, hospitalisation and death remains stellar (~20x), including against Omicron. There's also no moral failing in catching a respiratory virus. It's OK; it's life, which sucks at times. ++ It gives me no joy to announce the inevitability of SARS-CoV-2 becoming endemic, and in an ideal world I wished we could have avoided yet another respiratory virus circulating in the community (there are already ~200 including 4 HCoVs), but the world is just not always ideal. ++ I believe it is time to give in soon. Vaccine protection rates are as high as they may ever be in many places, and now we've got a couple of decent drugs. Pretending we remain in control, of sorts, is just becoming too costly. ++ To clarify, whilst I believe that we, as a society should try to extract ourselves out of the pandemic fairly soon, I wouldn't want to dictate anyone's behaviour, and I fully respect that some may need a bit longer before they feel ready to resume a 'normal' life. ++ It may be an understatement that this tread led to some heated replies. One misunderstanding is that I called for restrictions to be lifted right now. That wasn't my point. ++ Instead, I suggest we continue controlling covid case numbers over the coming months and keep proactive / reactive measures in place wherever they are needed to ensure as smooth as possible a transition into endemicity that will happen at different times in various settings. ++ Many also feel that a return to 'normal' would disproportionally affect the most vulnerable. I believe the picture is far more complex. It is true that pandemic mitigation measures, and lifting thereof, variably affect different people in society. ++ Though, overall, pandemic restrictions don't tend to have much of a negative impact on the healthy and wealthy. I know. I'm hardly incommoded by them, and personally won't benefit much for them to come to an end.

Show Me The Data
@txsalth2o

A virus so strong that it can get past 3 vaccines but can't get past your paw patrol mask.

9:47 PM · Dec 26, 2021 · Twitter Web App

Anish Koka
@anish_koka

Message I just received from a doctor lamenting the loss of nurses over the vaccine mandates :

" I would take an unvaxxed nurse over not having a nurse everyday "

10:26 AM · Dec 28, 2021 · Twitter for iPhone

Prof
@covidtweets

Things that OMGicron made the "experts" and the MSM realize (so far):

1. Cloth masks are a joke and surgical masks don't work either.

2. There are many hospitalizations who are not there because of COVID but just test + when screened.

3. Getting COVID is not a moral failure.

8:19 PM · Dec 25, 2021 · Twitter for iPhone

TEAM REALITY

Sarah Beth Burwick
@sarahbeth345

For fun, because I have a weird idea of fun, I just decided to see if I could find out how many people in Los Angeles County my age (37) have died of all causes since March 2020. According to the County Coroner records, there were 265.

How many were from covid, I wondered?

1:31 AM · Dec 29, 2021 · Twitter Web App

So I went through them all. Based on the records, 3 out of 265 died of covid-19. The medical examiner noted that two of the three were morbidly obese with heart conditions. The *vast* majority of deaths in 37yos were caused by alcohol drug abuse; fentanyl and methamphetamine. ++ The rest were primarily suicide and homicide. There were very few 37 year olds who died of illness unrelated to drugs or alcohol. But one after another seemed to be drugs. Meth. Fentanyl. Meth *and* fentanyl. Cocaine. Alcohol. Meth. Fentanyl. ++ All I can say for now is that maybe we are focusing on the wrong pandemic. ++ A postscript: to be clear, the death records reflected on the coroner's website do not represent all deaths in the county.

Ann Bauer
@annbauerwriter

The lesson of Covid isn't that climate change is a hoax. It's that today's Left will seize any complex problem and use it as a weapon to gain power, increase the wealth divide and provide luxury to the elite at the expense of everything you hold dear - without solving a thing.

11:55 AM · Dec 12, 2021 · Twitter Web App

Tracy Hoeg, MD, PhD
@TracyBethHoeg

🎖️ The research group at Oxford who compared myocarditis rates after covid to after the vax (nature.com/articles/s4159...) have just released a new analysis:
In males 16-39,
Myocarditis more common after Pfizer dose 2 & 3
&
Moderna dose 2 than after infection
medrxiv.org/content/10.110...

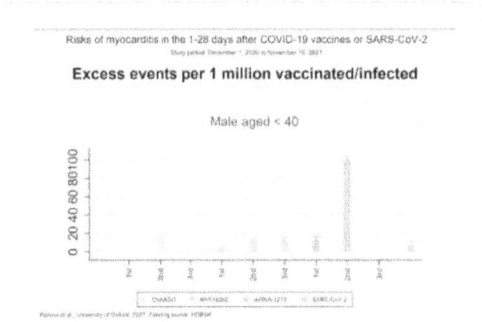

12:42 PM · Dec 26, 2021 · Twitter Web App

Emma Woodhouse
@EWoodhouse7

Masks and vaccines working well to stop "cases" in Chicago 😂

1:20 PM · Dec 28, 2021 · Twitter Web App

Jay Bhattacharya
@DrJBhattacharya

Bureaucrats who fund the careers of scientists should play no role whatsoever in setting pandemic policy. The conflict of interest created by this dual role has forced countless scientists to stay silent or risk losing their careers.

5:10 PM · Dec 18, 2021 · Twitter Web App

Eric
@The_OtherET

The Venn diagram of people who think we need to lock down again and the people who work in their jammies while they make others deliver their food and goods to them is a circle

8:51 AM · Dec 5, 2021 · Twitter for iPhone

TEAM REALITY

Emma Woodhouse
@EWoodhouse7

Dear Older People,

Society has "spared" almost 2 YEARS thinking about you - & barely a half a moment for children.

Destroying kids to assuage your fears is wrong.

I therefore reject the NYT's request to prolong this unholy pandemic response.

Love,
Me

IE MORNING NEWSLETTER

ovid's Risk to Older Adults

eady to give up on the pandemic? Spare a moment to think bout older people.

9:32 PM · Dec 26, 2021 · Twitter Web App

Eric
@The_OtherET

Have the mask fanatics ever considered for just one second that their overzealous exaggeration of masks' effectiveness is causing infected people with symptomatic illness to go out in public, incorrectly thinking that their loose cloth mask will protect others?

8:52 PM · Dec 27, 2021 · Twitter for iPhone

Jay Bhattacharya
@DrJBhattacharya

No scientist is in sole possession of the scientific truth. The truth emerges out of free discussion, experiment, and open minds. Why should the public -- which funds science -- not be allowed to witness the process?

9:05 AM · Dec 31, 2021 · Twitter Web App

Zac Bissonnette
@ZacBissonnette

If NYC passed a law that you can't post a selfie of yourself in line for a rapid test, the lines would be shorter.

8:39 PM · Dec 23, 2021 · Twitter for iPhone

Jenin Younes
@Leftylockdowns1

About half of my friends have Covid right now. Almost all of them vaccinated. Just putting that out there.

7:02 PM · Dec 21, 2021 · Twitter for iPhone

Tracy Hoeg, MD, PhD
@TracyBethHoeg

Many are asking for an update on our North Dakota masking study. Two K-12 public school districts in the same county: Cass, ND, both w/around 12,000 students
Active school cases in 2021 ⬇
FPS (●)= mask mandate
WF (●)= masks optional
FPS slightly higher staff vax rate
(1/2)

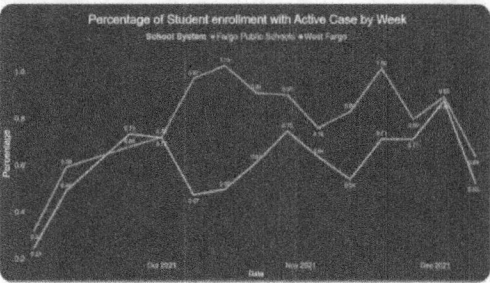

11:44 AM · Dec 12, 2021 · Twitter Web App

M_P
@Reroot_Flyover

I'm so happy for you Californians. With the new mask mandate you're finally two weeks from beating Covid!

6:07 PM · Dec 13, 2021 · Twitter Web App

Prof Francois Balloux
@BallouxFrancois

I'd be most surprised if come next year, anyone may claim they supported schools closures during the pandemic. I anticipate so much back-tracking and so much revisionism, the more savvy actors are already pivoting ,,,

7:16 PM · Dec 26, 2021 · Twitter Web App

Bethany S. Mandel
@bethanyshondark

Don't get masks you don't need them. jk you do. You can wear any kind of mask, what matters is you wear them. jk most of your masks don't work. Get vaccinated and you won't get sick. jk you still can. Get vaccinated and you can't spread it. jk you still can.

11:02 AM · Dec 28, 2021 · Twitter for iPhone

Eli Klein
@TheEliKlein

I have young healthy friends here in NYC that are Covid testing every day. One just txted me "How many tests have your taken this week?". I replied "None man lol".

12:36 PM · Dec 26, 2021 from Manhattan, NY · Twitter for iPhone

TEAM REALITY

Jay Bhattacharya
@DrJBhattacharya

So now I know what it feels like to be the subject of a propaganda attack by my own government. Discussion and engagement would have been a better path.

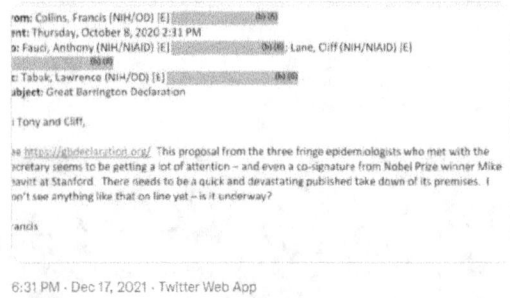

6:31 PM · Dec 17, 2021 · Twitter Web App

Prof Francois Balloux
@BallouxFrancois

I'd be most surprised if come next year, anyone may claim they supported schools closures during the pandemic. I anticipate so much back-tracking and so much revisionism, the more savvy actors are already pivoting ,,,

7:16 PM · Dec 26, 2021 · Twitter Web App

Erich Hartmann
@erichhartmann

As we watch the tens of millions of the Righteous catch COVID after double-masking for 22 months, injecting as many shots as legally possible, and embracing tyranny… we are now witnessing the inevitable mass realization:
All this societal damage was "completely unnecessary"
😤

9:05 AM · Dec 27, 2021 · Twitter Web App

M_P
@Reroot_Flyover

I still can't believe that Democrats decided they wanted to be the party of systematically excluding swaths of people from society.

Ooph.

4:25 PM · Dec 21, 2021 · Twitter Web App

Zac Bissonnette
@ZacBissonnette

I'm at a jazz club for a sold out 11PM show right now. No masks. All of these people are liberals.

And yet the NYT is still running stories, literally today, about when we might be able to stop cowering in fear of covid.

Many Democrats are terrified to admit they are over it.

10:44 PM · Dec 11, 2021 · Twitter for iPhone

Jennifer Sey
@JenniferSey

It feels a tad bit excessive to be in an official state of emergency while CA hosts the Super Bowl, the Rose Bowl, the Oscars and the governor goes on a book tour.

6:04 PM · Dec 12, 2021 · Twitter for iPhone

TEAM REALITY

JANUARY/FEBRUARY 2022

Jay Bhattacharya
@DrJBhattacharya

Institutionalized hypochondria is bad health policy.

6:01 PM · Jan 6, 2022 · Twitter Web App

TEAM REALITY

Jay Bhattacharya
@DrJBhattacharya

A noble lie is a lie.

6:02 PM · Jan 30, 2022 · Twitter Web App

District AI
@districtai

UK: back to normal

US and Biden: here are masks and "new normal"

This is political suicide for the Dems

4:55 PM · Jan 19, 2022 · Twitter Web App

AJ Kay
@AJKayWriter

Yesterday's oral arguments made it clear that we must include the US Supreme Court to the ever-growing list of once hallowed institutions that have been corroded (or had their corrosion exposed) by Covid hysteria.

6:43 AM · Jan 8, 2022 · Twitter for iPhone

Michael P Senger
@MichaelPSenger

Canadians set off fireworks last night as the truck convoy passed in protest against vaccine mandates and COVID restrictions. At over 70km long, Canada's convoy shatters the prior world record for longest truck convoy ever recorded.

From **RadioGenova**

2:33 PM · Jan 26, 2022 · Twitter Web App

Vinay Prasad, MD MPH
@VPrasadMDMPH

Wearing a mask from the door to the table, but at no point during the meal, in a crowded restaurant

Making kids wear masks outside at recess

Two examples of real policies that are metaphors for how the US handled COVID-19

3:06 PM · Jan 28, 2022 · Twitter Web App

ClownBasket
@ClownBasket

So let me get this straight, right now SCOTUS is deliberating on a mandate for a drug that the CEO of Pfizer states "...offers very limited protection if any."

10:07 PM · Jan 10, 2022 · Twitter for iPhone

Jay Bhattacharya
@DrJBhattacharya

Lockdowners think that lockdowns must work because they stop the dangerous act of people meeting with other people. They forget that lockdowns are a luxury of the laptop class, and barring draconian overreach that kills, essential human interactions will continue regardless.

2:24 PM · Feb 5, 2022 · Twitter Web App

Jennifer Cabrera
@jhaskinscabrera

I hope COVID teaches the people of America that we must never again let government impose restrictions for a virus until a vaccine is ready. And obviously, we should stop complying immediately.

As many of us predicted: once you give up freedom, it's not easily regained.

10:14 AM · Jan 20, 2022 · Twitter Web App

David M
@ComradeDoom1

the ACLU deciding that every child has an inalienable right to be forced to wear a mask has got to be some kind of watershed moment in American legal history.

2:26 AM · Feb 16, 2022 · Twitter Web App

Aaron Kheriaty, MD
@akheriaty

Gonna be a lot of jobs for cardiologists in the next several decades.

4:41 PM · Jan 27, 2022 · Twitter Web App

Brumby
@the_brumby

In the last ~45 days millions of Americans have had a cold, taken a test to find out the cold was caused by Covid, realized there is absolutely no benefit to themselves or society of having confirmed such information, & perhaps r beginning to wonder what the hell we've been doing

4:34 PM · Jan 20, 2022 · Twitter Web App

Don Wolt
@tlowdon

UK HSA data shows infection rate growth correlates with vaccination rate for age cohorts >50. The more highly vaxxed an age cohort is - whether fully vaxxed or boosted - the faster the infection rate growth. Infectn rate growth in the unvaxxed is fairly consistent across cohorts.

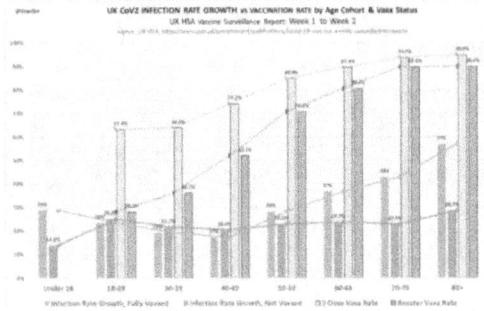

2:20 PM · Jan 16, 2022 · Twitter Web App

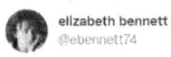
elizabeth bennett
@ebennett74

That moment when you smile at another unmasked adult in a store before you realize there is a masked child trailing them 🤦 (I retract my smile at you, sir)

5:37 PM · Jan 26, 2022 · Twitter for iPhone

Justin Hart
@justin_hart

I encourage all Democrats to run on masking kids, expelling people from their jobs, quarantining healthy people and mandating more jabs. Please, please run on those issues.

8:51 AM · Feb 15, 2022 · Twitter for iPhone

Jay Bhattacharya
@DrJBhattacharya

The public's capacity for compliance with public health mandates and advice is finite. Public health energy spent on low value nonsense, like vax mandates for the young and masking toddlers, has enormous opportunity costs paid in lives of the unprotected vulnerable.

9:10 AM · Feb 13, 2022 · Twitter Web App

Phil Kerpen ✓
@kerpen

This is just absolutely astonishing. "100,000 children in serious condition," per Sotomayor. Where do these people obtain their misinformation? The current national pediatric COVID census per HHS is 3,342. Many/most incidental.

10:55 AM · Jan 7, 2022 · Twitter Web App

Dr Clare Craig (not one of her impersonators)
@ClareCraigPath

Yesterday I was in the Royal Courts of Justice hearing the child vaccination case.

Two things struck me:
1. The people at the ONS have lost their moral compass
2. Numbers don't cut it

5:21 AM · Jan 14, 2022 · Twitter Web App

Victoria Fox
@drvictoriafox

As a scientist I used to think pandemics ended when the population became immune to the pathogen.

I now think if fully vaxed & boosted colleges are requiring their already low risk students to go remote, test, mask & distance - this pandemic will only end through the courts.

6:46 PM · Jan 11, 2022 · Twitter Web App

Stefan Baral ✓
@sdbaral

My experience having to wear N95s in outbreak settings only reinforces my perspective that making kids wear them all day long in school is ridiculous.

10:24 AM · Jan 1, 2022 · Twitter Web App

Abir Ballan
@abirballan

The president of the South African Medical Association, Angelique Coetzee, "In #SouthAfrica, it [#Omicron] is a mild disease, but it's a serious, serious, serious illness in #Europe, because politicians want me to say it"

4:25 AM · Jan 24, 2022 · Twitter for iPhone

Don Wolt
@tlowdon

Say you're the CDC Director & several SCOTUS Justices make wildly inaccurate & misinformed statements overstating the impact of both COVID & its vaccines. Do you (A) quickly correct the record with the public or (B) remain quiet knowing the misinformation serves your purposes?

5:17 PM · Jan 8, 2022 · Twitter Web App

TEAM REALITY

Josh Stevenson
@ifihadastick

There are some folks attacking a group of incredible, empathetic, accomplished, data & research obsessed, kind and humble people who are genuinely telling parents the truth about the harms of Covid vs the harms of severe restrictions upon children, and equipping/ educating

1/

12:00 AM · Jan 26, 2022 · Twitter Web App

them to provide much needed normalcy for their kids who are suffering. One of the disingenuous things they are doing is trying to lie about the normal burden of respiratory disease. They are lying about flu- using data from when Flu was not circulating to tell you Covid ++ is worse for children. This is absolutely false and the data provided at https://urgencyofnormal.com is well resources and citations are provided. These experts reputation and life they invested in their careers is not something they would give up to suddenly become part ++ Of a "disinformation" campaign. The accusation is laughable. And it shows how toxic and hateful those who demand that children be restricted are behaving. But, to the data. Yes- would it surprise you to realize that though yes, kids can get and die from Covid, that this ++ reality is what we have experienced with many respiratory pathogens every year? And that mortality during SARS-COV2, as well as hospital burden, is at or possible below what it typically is every year? ++ What these players are trying to portray is a false crisis that is particular to children and that both Covid is highly fatal (it's not, it's .0013 IFR - to this group as well as "Long Covid" is a major risk (also false https://journalofinfection.com/article/S0163-4453(21)00555-7/fulltext…) ++ What these courageous people are trying to do is to re-calibrate peoples perception of risk, by aligning with the data. That just happens to mean being rational and empathetic because the data says we can. ++ If you weren't taking extreme measures for your kids during past respiratory virus seasons, you don't need to do that for Sars-Cov2. It's not going away, and it's also not the only pathogen out there. This is good news, not bad.

Jay Bhattacharya
@DrJBhattacharya

So many good people sacrificed so much to comply with the failed Covid policies. I am grateful to live in a country with so many public-spirited people. It is the architects of lockdowns & the panic-mongers whose thinking needs repudiation so their mistakes are never repeated.

12:06 AM · Jan 25, 2022 · Twitter Web App

David M
@ComradeDoom1

to be a good leftist in 2022 means that you're capable of sincerely believing that parents only object to having their kids wear a mask all day in perpetuity because they're being paid by the Koch brothers.

11:54 PM · Feb 12, 2022 · Twitter Web App

District AI
@districtai

Truckers and Joe Rogan finally broke the left.

Not what I had on my bingo card for 2022

6:03 AM · Jan 31, 2022 · Twitter Web App

Vinay Prasad, MD MPH
@VPrasadMDMPH

When people say "we are never returning to normal" they just mean they aren't. Everyone else will soon.

9:41 PM · Jan 26, 2022 · Twitter Web App

Eli Klein
@TheEliKlein

I am so embarrassed that more NYC business owners aren't standing up against the Covid vaccine passport. As a City we've fallen so far. Who the fuck are you to require a 5 year old that just had Omicron get vaccinated against the disease they already have excellent immunity to?

6:34 PM · Jan 29, 2022 from Manhattan, NY · Twitter for iPhone

NYC Angry Mom
@angrybklynmom

Very few of my old friends/family will engage with me on this topic, but I wish I could understand how ppl who spent 4+ yrs obsessed w/ Trump's authoritarianism are essentially silent right now in the face of the largest loss of freedom we have ever experienced in our lifetimes.

7:52 AM · Feb 22, 2022 · Twitter for Android

Do they feel it is justified in the name of safety? Are they not concerned because this loss of freedom is primarily being driven by the political party that they trust? Do they just assume these freedoms will be restored when the emergency ends? ++ Have they bothered to consider what metrics drive the end of the emergency? Have they looked at how other states/countries are handling COVID without the accompanying loss of freedoms? ++ Have they evaluated the overall results in areas with heavy-handed restrictions vs areas without, including other countries with similar political structures? Have they considered the overall impact to society, and the causes they care about? ++ Do they have empathy for the millions of people who were irreparably harmed by these heavy-handed restrictions, or does empathy only flow toward people impacted by the virus? Do they still believe that viral spread is a moral issue, after 2 years? ++ I just want to understand what can be going on in people's minds when they watch Western democracies across the globe implement some of the most draconian, unfathomable violations of both human rights and civil liberties that we have ever seen. ++ Is this apathy ultimately rooted in a simple exaggeration of COVID risk in the young? Do people understand the haphazard manner in which COVID hospitalizations and deaths are counted? Do they understand the age stratification of risk? ++ Does it not concern people that the youngest people in the country are paying the steepest price in terms of loss from restrictions for a virus that is incredibly low risk to the vast majority of them? And that this is continuing far past widespread vaccine availability? ++ I collected dozens of articles from Australia and New Zealand for the past 2 years, and even now, when I read some of them, I am simply amazed that they were able to harm citizens in this manner and that they had support from their own population and from ours to do this. ++ And now we see Canada invoking a brand new state of emergency that has only been invoked during wartime, so that they can freeze the bank accounts of people who donated small sums of money to a grassroots, non-violent protest. ++ And nobody cares. I have not seen a single person from my pre COVID life expressing concern about this. Are they not aware that Trudeau decided to extend this "emergency" for a few weeks, "just in case"? Why? Can you imagine Trump doing that? ++ It doesn't seem to bother people that US college students are being treated like prisoners, despite complying with vaccination & booster mandates, even when the data on boosters being necessary for this age group were weak at best. Why doesn't this bother you? ++ Or do you simply reduce everything to simplistic narratives like "anti-masker" or "muh freedumb" so you can sleep soundly at night, with the righteous belief that you are one of the "good guys" who "cares about others"? ++ Even if you don't actually care about what's happening around you right now, what is really shocking is that you don't seem to care about how the extremism we are living through will simply result in more extremism, but from the other side. You're enabling this. ++

TEAM REALITY

We are not going to live through years of heavy-handed government restricting the very basics of our freedoms, and causing irreparable damage to the population in the name of a virus they ultimately could not control, without some type of political upheaval. ++ If you're not speaking out against the complete and total insanity that we have been subjected to for 2 years, consider yourself complicit when the population ultimately revolts & elects people who are far more extreme in the other direction.

ClownBasket
@ClownBasket

My kid's mask allegedly protects you.

But my kid's mask actually harms them.

Who's the selfish one?

7:25 PM · Feb 12, 2022 · Twitter for iPhone

Woke Zombie
@AWokeZombie

My Son asked me today. "Why do kids who arent vaccinated have to quarantine if the vaccinated kids are still getting and spreading covid?

He's 10. He has more sense than pretty much every single person in public health.

And that, in a question, is whats wrong with all of this.

10:16 PM · Feb 18, 2022 · Twitter Web App

Jay Bhattacharya
@DrJBhattacharya

Pres. Biden got several things wrong in this clip:
1. This is not just an epidemic of the unvaccinated
2. The most important step to protect kids is to let them live normal lives
3. Surrounding toddlers with unvaccinated toddlers is perfectly ok and good for their development

7:03 PM · Jan 4, 2022 · Twitter Web App

ClownBasket
@ClownBasket

Replying to @aprylmarie

It's easy to mistaken "kids are resilient" with Kids Have No Agency.

10:57 PM · Jan 30, 2022 · Twitter for iPhone

Prof Francois Balloux
@BallouxFrancois

"Kids are resilient" is one the most idiotic statements I've ever come across. It's also ironic that it is very generally uttered by adults who are anything but.

8:54 PM · Jan 4, 2022 · Twitter Web App

Ian Miller
@ianmSC

Cases in the United States are now down 86.4% since the Surgeon General said the Omicron wave hadn't peaked and the next few weeks would be "tough"

The most important thing to remember about COVID policy is that The Experts™ have absolutely no idea what they're talking about

1:59 PM · Feb 21, 2022 · Twitter for Mac

Hold2
@Hold2LLC

And please tell me how this is not a data crime when @nycHealthy switches a NET 115,310 Cases from Unvaxxed to Vaxxed with no announcement by anyone?

Not Commissioner Chokshi @NYCHealthCommr , @MarkLevineNYC , @zeynep , @sailorrooscout , @StevenTDennis, @guypbenson , or anyone.

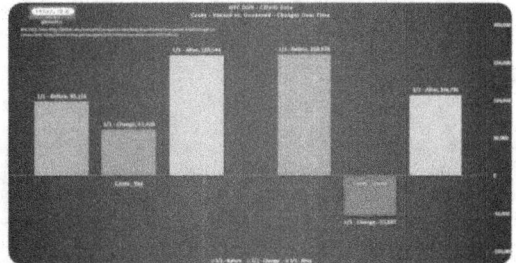

9:37 PM · Feb 11, 2022 · Twitter Web App

TEAM REALITY

Matt, Pre-School Diploma 😀
@statomattic

Those of us who have resisted the government/media covid narrative have been called many things over the past two years, most of them cruel and false. I thought it time to put on paper what we really ARE. So, with the help of some of my friends, here you go… 🧱

9:07 PM · Jan 11, 2022 · Twitter for Android

You may not realize it, but we live near you. We are your neighbors. We are your coworkers. We are your friends who have been quiet to preserve friendships. We are men and women. Many of us are parents. We are all races. We are all faiths. We come from all political backgrounds. ++ Based on the overwhelming preponderance of data from around the globe since early 2020, as well as the preponderance of scientific literature prior to 2020, we agree that lockdown policies and mask mandates are ineffective instruments to mitigate a virus. ++ Their costs far outweigh their negligible benefits…. ++ We are grateful for the benefits of modern medicine. We also believe the benefits of any medicine must be tracked empirically, weighed against costs for each individual. Accordingly, we believe all personal medical decisions should be private, and neither mandated nor coerced. ++ We are indebted to science, the greatest method of inquiry human beings have devised to help explain how the world around us works. ++ In that spirit, claiming the mantle of "The Science" in an effort to STIFLE inquiry and debate to serve political ends is NOT science. It is a cynical bastardization of science, a mockery of its good name. ++ As we believe each human life is precious, we are aware that the experiences that become memories give life its richness. Because no person can know when life will end or when physical vitality will erode… ++ …we believe firmly that the days of our lives are not meant to be wasted. Adopting the supposed timeline of a virus is therefore not a mere inconvenience, but a forfeiture of days we cannot recover. ++ We believe it is morally repugnant to sacrifice the priceless, formative experiences of children in an effort of questionable worth to arguably add limited time frames to the feeble ends of elderly lives. ++ We believe physical, in-person interaction is essential to the human experience. So-called "virtual" communication is not an equal substitute. ++ We believe all occupations are essential. ++ We believe the human face is an essential, irrepressible aspect of humanity. ++ We believe population-wide mask mandates and lockdowns are human rights violations. ++ We are open to being wrong. We welcome robust, civil debate. ++ We are Team Reality. If you would like to get connected with like-minded people in your state, please reach out to the person listed for your location on the attached thread. We are welcoming, friendly, curious, and would love to meet you.

Sarah Beth Burwick
@sarahbeth345

I know many people who want mask mandates to continue, and they have one thing in common: they work from home. Wearing a mask for eight hours a day is desirable to them, as long as they are not the ones who have to do it.

10:52 AM · Feb 21, 2022 · Twitter for iPhone

Kyle Lamb
@kylamb8

The dam is broke. A barrage of mandates are ending.

But it shows how this is all political as it's convenient how all these blue states decided to drop the mandates at the same time as if a memo went out from DNC headquarters.

6:42 PM · Feb 7, 2022 · Twitter Web App

TEAM REALITY

Sarah Beth Burwick
@sarahbeth345

"Covid-minimizer" is another meaningless label people slap on you when they know they're wrong. Demanding nuanced policies based on risk assessment informed by data, while balancing collateral harms ≠ minimizing.

The fear monger covid maximizers are who we need to worry about.

10:02 AM · Feb 26, 2022 · Twitter for iPhone

Emma Woodhouse
@EWoodhouse7

No matter what the updated CDC mask guidelines say today, never forget the governors, superintendents, school board members, & Teacher Union leaders who were fighting til the bitter end to force-cover the faces of toddlers thru teens.

10:25 AM · Feb 25, 2022 · Twitter for iPhone

Emma Woodhouse
@EWoodhouse7

Monument to U.S. Covid-response policy & public-health messaging failures, right here.

Two years & trillions of dollars wasted on a fool's errand.

11:28 AM · Feb 4, 2022 · Twitter for iPhone

Bethany S. Mandel
@bethanyshondark

COVIDians in 2020: Look at how bruised and bloody n95s make the faces of healthcare workers.
COVIDians in 2022: Let's put n95s on 3 year olds.

8:46 AM · Jan 18, 2022 · Twitter for iPhone

Emily Burns DMs welcome #TeamReality
@Emily_Burns_V

1/
Looking at 10 MOST-VAXED states, and 10 LEAST-VAXED states, we see that comparing Year-of-Year % change, there is virtually NO DIFFERENCE...

Except that NOW, the 10 most-vaxed have nearly 400% HIGHER cases than last year, where 10 LEAST-vaxed, "only" up 112%.

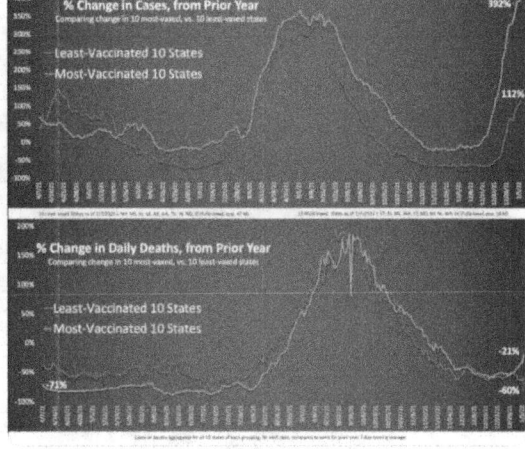

11:07 AM · Jan 11, 2022 · Twitter Web App

Jay Bhattacharya
@DrJBhattacharya

At this point in the epidemic, the main purpose of contact tracing is to keep kids out of school, make sure hospitals & businesses are short-staffed, and impose lockdown by stealth.

9:09 PM · Jan 8, 2022 · Twitter Web App

TEAM REALITY

Eric @The_OtherET

Been seeing a lot of governors crediting their mask mandates for "defeating" the Omicron wave. Thought I'd plot COVID cases in states with mask mandates vs. states without them, and, well...

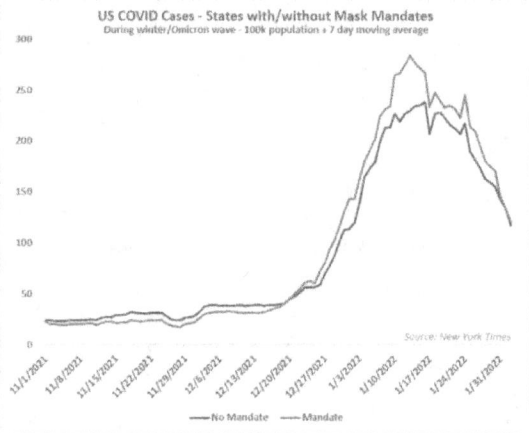

4:29 PM · Feb 3, 2022 · Twitter Web App

BOUTROS 木 @boutros555

The thing about the N95 masks is that the damn things have already been mandated in Bavaria & were as ineffective as the cloth variety.

Universal N95 use is not PH guidance. It's "sciency" sleight-of-hand in a desperate attempt to preserve a flailing bureaucracy's public image.

8:17 PM · Jan 14, 2022 · Twitter for iPad

220 Retweets   7 Quote Tweets   1,199 Likes

Jay Bhattacharya @DrJBhattacharya

The Fauci/Collins playbook to create a false impression of scientific consensus on COVID policy (used on lab-leak, lockdowns & early treatment):
1. Call scientists who disagree "fringe"
2. Deploy big tech misinformation hordes to suppress opposing thoughts
[1/2]

8:00 PM · Feb 2, 2022 · Twitter Web App

Jay Bhattacharya @DrJBhattacharya

3. Deploy press propagandists & scientist allies to smear and takedown opponents
4. Reward allies with large grants
[2/2]

8:00 PM · Feb 2, 2022 · Twitter Web App

Ann Bauer @annbauerwriter

"SHOW ME YOUR PAPERS" has never, in all of recorded history, been the right side.

This time is not different.

9:40 AM · Jan 21, 2022 · Twitter Web App

Bethany S. Mandel @bethanyshondark

If nobody minds wearing a mask, why is nobody wearing them at the Super Bowl

9:21 PM · Feb 13, 2022 · Twitter for iPhone

Jenin Younes @Leftylockdowns1

I just learned that Jacinda Arden is delivering Harvard's commencement speech, makes a lot of sense considering that they share the values of intolerance and authoritarianism. Maybe Justin Trudeau will deliver Yale's.

4:07 PM · Feb 15, 2022 · Twitter Web App

Vinay Prasad, MD MPH @VPrasadMDMPH

Every organization mandating boosters in young healthy people knowing the data on omicron's breakthrough rate despite boosting is making a big error.

It is ethically bankrupt, it is scientifically bankrupt, and it is going to cause serious damage to future vaccination efforts.

4:43 PM · Jan 28, 2022 · Twitter Web App

Wes Pegden @WesPegden

Omicron affects everybody:

Children: No school
College kids: No lectures (at some places: no life!)
Adults: Wear a mask from the hostess stand to your table / the bar.

We're all in this together.

9:55 PM · Jan 5, 2022 · Twitter Web App

M_P @Reroot_Flyover

Thank you @TheDemocrats for coordinating your Covid pivot and making it abundantly clear that politics was in the driver seat all along and not "science".

6:40 PM · Feb 7, 2022 · Twitter Web App

TEAM REALITY

3. THE AFTERMATH

Stefan Baral ✓
@sdbaral

Replying to @ajlamesa

It's an old lesson in public health and medicine--it is easier to scare folks than unscare them.

10:24 AM · Mar 19, 2022 · Twitter Web App

Jay Bhattacharya
@DrJBhattacharya

Now that the politics favor the pandemic ending, the pandemic is ending.

It was a coalition of regular people (working-class, poor, moms) who forced the shift.

Next, they will hold accountable the people who pushed the lockdowns to answer for the destruction they caused.

5:35 PM · Mar 11, 2022 · Twitter Web App

Whatever your views are on the severity of the disease caused by SARS-COV2 itself, there is no denying that the pandemic was a major event which caused a significant disruption in our way of life. I always hated the term "new normal" and I still reject it with passion. However, it would be naïve to deny that the pandemic will have significant repercussions for the years to come. Here, I will summarize those in three areas: Loss of trust in institutions, damage caused by NPIs, and political ripple effects.

First, if there is one thing that Team Reality and Team Apocalypse (this is the first time I am using this name in this book, but it is what the doomers were often called on Twitter) agree on, it is that our COVID response was a major screw-up. The doomers are upset that we did not follow zero COVID, while Team Reality argues that almost everything we did were unnecessary and caused more harm than good. Inevitably, this has led to a massive distrust in the institutions from both sides. More importantly, the general public, who would identify with neither team, has seen how clueless our institutions were. This will have profound impact on how people are going to view our institutions in the years to come, including our scientific establishment. For example, I for one had never questioned the science supporting climate change before the pandemic. Now I am at a stage where I say, "I need to see the evidence for myself". This is not necessarily a bad thing, but excessive skepticism is not very healthy either. We should be able to trust our scientists or bureaucrats who are running our most important institutions, CDC being one of them. I pray that we do not face another major crisis for at least the next ten years, so that maybe by then some of the trust that was lost during the pandemic will be regained. (I pick up my phone to check Twitter as I am writing this, and the first thing I see is a notification which shows that someone quoted one of my tweets in which I criticized an "expert" for claiming (obviously falsely) that "COVID is a worse disease, in children, than polio". The tweet which quoted mine says "Your healthcare experts have utterly FAILED the American people. We will NEVER trust you again.")

Second, the massive mistakes made in the name of "flattening the curve", "stopping the spread", "slowing the spread", "controlling COVID", or whatever it will be called in 2029, will have drastic ripple effects for years to come. The damage caused by the insane policies which became the default in 2020-2021 will continue to haunt us for many years to come. The most consequential of them all, school closures, which have already widened the gap between kids from disadvantaged families and those from more affluent families, will leave a lifetime mark on those kids' well-being. The fact that many used "kids are resilient" to justify these cruel policies deserves a book in and of itself. Those kids will see reduced earnings throughout their lives, lower quality of life, and even reduced life expectancy because of what we did to them. I have tweeted this a couple of times: I really hope they realize what was done to them and make us pay for our crimes when they have the power and we are vulnerable...

Missed childhood vaccinations, missed cancer screenings, missed annual physicals, postponed treatments for normally treatable conditions making them worse, a more sedentary lifestyle caused by "staying home" for months, a more sedentary lifestyle caused by the increase in remote work, and perhaps the long-term adverse events from boosters given to people who didn't really need them will all contribute to an overall worsening of our collective physical health and excess deaths for the years to come. The total life years lost from these combined will likely dwarf the life years lost because of COVID. Our collective mental health deterioration will also continue to have consequences for a while, leading to a worse life outlook for many people who would otherwise be well-functioning members of society.

Third, the pandemic was a wake-up call for many people which made them realize how precious civil liberties were, how we were taking them for granted, and how they could be taken away in an instant when those in power think the conditions justify it. There will be a political reshuffling because of what politicians chose to do during the pandemic. I believe this will either benefit Republicans in the long term or will result in the emergence of a major third party mainly supported by people who now trust the political elite even less. I am not knowledgeable enough on US politics to have more specific predictions on this, but I am highly confident that there will be some major consequences in the political landscape, that we have begun to see in the elections in 2021 and will see more of in the coming midterms.

I left several empty pages at the end of this book, because I know this is not even close to fully capturing the full extent of the damage we caused. There will be many more unintended consequences to our collective stupidity. I left those pages so that you can add your own notes over the years as we come to realize how disastrous our handling of COVID was. Then leave this book to your kids so they know you were on the side of reason, but it was hard to overcome organized insanity...

TEAM REALITY

# TEAM REALITY

TEAM REALITY

REFERENCES

1. https://www.euroweeklynews.com/2020/04/03/the-wuhan-connection-does-new-evidence-reveal-why-northern-italy-became-the-nations-coronavirus-epicenter/
2. https://www.ncbi.nlm.nih.gov/pmc/articles/PMC7430577/
3. https://www.forbes.com/sites/geoffwhitmore/2020/10/19/when-did-president-trump-ban-travel-from-china-and-can-you-travel-to-china-now/?sh=6256ce2f7484
4. https://www1.nyc.gov/office-of-the-mayor/news/079-20/mayor-de-blasio-speaker-johnson-queens-chamber-commerce-encourage-new-yorkers-visit/#/0
5. https://www.nbcbayarea.com/news/local/nancy-pelosi-visits-san-franciscos-chinatown/2240247/
6. https://www.foxnews.com/politics/dems-media-change-tune-trump-attacks-coronavirus-china-travel-ban
7. https://www.imperial.ac.uk/media/imperial-college/medicine/sph/ide/gida-fellowships/Imperial-College-COVID19-NPI-modelling-16-03-2020.pdf